ACCLAIM FOR JOY WILLIAMS AND *ILL NATURE*

"Enchanting and explosive."

—*The Washington Post Book World*

"Glows with fire-and-brimstone passion."

—*The Boston Globe*

"Troubling, fascinating. . . . Pointedly detailed, painfully entertaining. . . . It adds up to a jeremiad classically American and on-the-edge uncanny."

—*The Nation*

"Rise[s] above the din of dreary environmental writing and smack[s] us in the face with the sorry state of our natural affairs."

—*San Francisco Chronicle*

"Joy Williams brings her fierce compassion to bear on subjects she holds dear: the vanishing Florida Everglades, the miracle of animals in our lives, why she writes. Her words are tonic, wholly original; she is a writer to reckon with."

—*O: The Oprah Magazine*

"Joy Williams' essays . . . manage to articulate with wit, elegance, intelligence, and appropriate disdain, the enterprise in which we are all implicated."

—W. S. Merwin

ILL NATURE

ILL NATURE

*Rants and Reflections on Humanity
and Other Animals*

JOY WILLIAMS

Guilford, Connecticut

An imprint of Rowman & Littlefield

Distributed by NATIONAL BOOK NETWORK

Copyright © 2001 by Joy Williams
Lyons Paperback Edition, 2016

British Library Cataloguing in Publication Information Available

The Library of Congress has previously catalogued an earlier edition of this book as follows:

Williams, Joy, 1944–
Ill nature : rants and reflections on humanity and other animals / Joy Williams
p.   cm.
Originally published: New York : Lyons Press, © 2001.
ISBN: 0-375-71363-8 (trade paper)
Nature—Effect of human beings on. 2. Environmental ethics. 3. Environmental degradation—Moral and ethical aspects. 1. Title
GF75.W56 2002
179'.1—dc21

2002020114

ISBN 978-0-7627-9650-2 (pbk.)
ISBN 978-1-4930-2371-4 (e-book)

∞™ The paper used in this publication meets the minimum requirements of American National Standard for Information Sciences—Permanence of Paper for Printed Library Materials, ANSI/NISO Z39.48-1992.

To Richard Ford

CONTENTS

ACKNOWLEDGMENTS

The essays, some in different form, first appeared in magazines, as follows:

"Save the Whales, Screw the Shrimp" and "The Killing Game" appeared in *Esquire* magazine. "Sharks" appeared in a much longer form as "The Bad News About Sharks" in *Esquire*. "Electric Chair" appeared in a much longer form as "The Angel of Death Row" in *Esquire*.

"The Case Against Babies" and "Hawk" appeared in *Granta*.

"Safariland" appeared in *Outside* magazine. "Neverglades" appeared in a much longer form in *Outside*.

"The Animal People" appeared as "The Inhumanity of the Animal People" in *Harpers* magazine.

"Suicide" appeared in a much longer form as "The Love Song of Wendy O. Williams" in *Spin* magazine.

"Coral Castle" appeared in *Nest*.

"Cabin Cabin" appeared in a much longer form in *Joe* magazine.

"One Acre" appeared in *Harper's*.

"Wildebeest" appeared in the *Black Warrior Review*.

"Cats" appeared in *River City Magazine* under the title "Cruiseship."

"Why I Write" first appeared in *Oxford American* but was originally commissioned for the anthology *Why I Write* edited by Will Blythe (Little, Brown), where it subsequently appeared.

Save the Whales, Screw the Shrimp

I DON'T WANT TO TALK ABOUT *ME*, OF COURSE, BUT IT SEEMS AS
though far too much attention has been lavished on you lately—
that your greed and vanities and quest for self-fulfillment have
been catered to far too much. You just want and want and want.
You believe in yourself excessively. You don't believe in Nature
anymore. It's too isolated from you. You've abstracted it. It's so
messy and damaged and sad. Your eyes glaze as you travel life's
highway past all the crushed animals and the Big Gulp cups.
You don't even take pleasure in looking at nature photographs
these days. Oh, they can be just as pretty as always, but don't
they make you feel increasingly ... anxious? Filled with more
trepidation than peace? So what's the point? You see the picture
of the baby condor or the panda munching on a bamboo shoot,
and your heart just sinks, doesn't it? A picture of a poor old sea
turtle with barnacles on her back, all ancient and exhausted,
depositing her five gallons of doomed eggs in the sand hardly
fills you with joy, because you realize, quite rightly, that just outside the frame falls the shadow of the condo. What's cropped
from the shot of ocean waves crashing on a pristine shore is the
plastics plant, and just beyond the dunes lies a parking lot. Hidden from immediate view in the butterfly-bright meadow, in
the dusky thicket, in the oak and holly wood, are the surveyors'

stakes, for someone wants to build a mall exactly there—some gas stations and supermarkets, some pizza and video shops, a health club, maybe a bulimia treatment center. Those lovely pictures of leopards and herons and wild rivers—well, you just know they're going to be accompanied by a text that will serve only to bring you down. You don't want to think about it! It's all so uncool. And you don't want to feel guilty either. Guilt is uncool. Regret maybe you'll consider. *Maybe.* Regret is a possibility, but don't push me, you say. Nature photographs have become something of a problem, along with almost everything else. Even though they leave the bad stuff out—maybe because you know they're leaving all the bad stuff out—such pictures are making you increasingly aware that you're a little too late for Nature. Do you feel that? Twenty years too late? Maybe only ten? Not way too late, just a little too late? Well, it appears that you are. And since you are, you've decided you're just not going to attend this particular party.

• • •

Pascal said that it is easier to endure death without thinking about it than to endure the thought of death without dying. This is how you manage to dance the strange dance with that grim partner, nuclear annihilation. When the US Army notified Winston Churchill that the first A-bomb had been detonated in New Mexico, it chose the code phrase BABIES SATISFACTORILY BORN. So you entered the age of irony, and the strange double life you've been leading with the world ever since. Joyce Carol Oates suggests that the reason writers—real writers, one assumes—don't write about Nature is that it lacks a sense of humor and registers no irony. It just doesn't seem to be of the times—these slick, sleek, knowing, objective, indulgent times. And the word *environment*. Such a bloodless word. A flat-footed word with a shrunken heart. A word increasingly disengaged from its association with the natural world. Urban planners, industrialists, economists, developers use it. It's a lost word, really. A cold word, mechanistic, suited strangely to the coldness generally felt toward Nature. It's their word now. You don't mind giving it up. As for *environmentalist*, that's one that can really bring on the yawns, for you've tamed and tidied it, neutered it quite nicely. An environmentalist must be calm, rational, reasonable, and willing to compromise; otherwise, you won't listen to him. Still, his beliefs are *opinions* only, for this is the age of radical subjectivism. Some people might prefer a Just for Feet store to open space, and they shouldn't be castigated for it. All beliefs and desires and needs are pretty much equally valid. The speculator has just as much right to that open space as the swallow, and the consumer has the most rights of all. Experts and computer models, to say nothing of lawsuits, can hold up environmental checks and reform for decades. The Environmental Protection Agency protects us by finding "acceptable levels of harm" from pollutants and then issuing rules allowing industry to pollute to those levels. Any other approach would place limits on economic growth. Limits on economic growth! What a witchy notion! The EPA can't keep abreast of progress and its unintended consequences. They're drowning in science. Whenever they do lumber into action and ban a weed killer, say (and you do love your weed killers—you particularly hate to see the more popular ones singled out), they have to pay all disposal costs and compensate the manufacturers for the market value of the chemicals they still have in stock.

That seems . . . that seems only fair, you say. Financial loss is a serious matter. And think of the farmers when a particular effective herbicide or pesticide is banned. They could be driven right out of business.

Farmers grow way too much stuff anyway. Federal farm policy, which subsidizes overproduction, encourages bigger and bigger farms and fewer and fewer farmers. The largest farms don't produce food at all, they grow feed. One third of the wheat, three quarters of the corn, and almost all of the soybeans are used for feed. You get cheap hamburgers; the agribusiness moguls get immense profits. Subsidized crops are grown with subsidized water created by turning rivers great and small into a plumbing system of dams and irrigation ditches. Rivers have become conduits. Wetlands are increasingly being referred to as *filtering systems*—things deigned useful because of their ability to absorb urban runoff, oil from roads, et cetera.

We know that. We've known that for years about farmers. We know a lot these days. We're very well informed. If farmers aren't allowed to make a profit by growing surplus crops, they'll have to sell their land to developers, who'll turn all that arable land into office parks. Arable land isn't Nature anyway, and besides, we like those office parks and shopping plazas, with their monster super-markets open twenty-four hours a day and aisle after aisle after aisle of products. It's fun. Products are fun.

• • •

Farmers like their poisons, but ranchers like them even more. There are well-funded federal programs like the Agriculture

Department's "Animal Damage Control Unit," which, responding to public discomfort about its agenda, decided recently to change its name to the euphemistic Wildlife Services. Wildlife Services poisons, shoots, and traps thousands of animals each year. Servicing diligently, it kills bobcats, foxes, black bears, mountain lions, rabbits, badgers, countless birds—all to make this great land safe for the string bean and the corn, the sheep and the cow, even though you're not consuming as much cow these days. A burger now and then, but burgers are hardly cows at all, you feel. They're not all our cows, in any case, for some burger matter is imported. There's a bit of Central American burger matter in your bun. Which is contributing to the conversion of tropical rain forest into cow pasture. Even so, you're getting away from meat these days. You're eschewing cow. It's seafood you love, shrimp most of all. And when you love something, it had better watch out, because you have a tendency to love it to death. Shrimp, shrimp, shrimp. It's more common on menus than chicken. In the wilds of Ohio, far, far from watery shores, four out of the six entrees on a menu will be shrimp something-or-other, available, for a modest sum. Everywhere, it's all the shrimp you can eat or all you care to eat, for sometimes you just don't feel like eating all you can. You are intensively harvesting shrimp. Soon there won't be any left, and then you can stop. Shrimpers put out these big nets, and in these nets, for each pound of shrimp, they catch more than ten times that amount of fish, turtles, and dolphins. These, quite the worse for wear, are dumped back in. There is an object called TED (Turtle Excluder Device) that would save thousands of turtles and some dolphins from dying in the net, but shrimpers are loath to use TEDs, as they argue it would cut the size of their shrimp catch.

We've heard about TED, you say.

At Kiawah Island, off the coast of South Carolina, visitors go out on Jeep "safaris" through the part of the island that hasn't been developed yet. ("Wherever you see trees," the guide says, "it's actually a lot.") The visitors (i.e., potential buyers) drive their own Jeeps, and the guide talks to them by radio. Kiawah has nice beaches, and the guide talks about turtles. When he mentions the shrimpers' role in the decline of the turtle, the shrimpers, who share the same frequency, scream at him. Shrimpers and most commercial fishermen (many of them working with drift and gill nets anywhere from six to thirty miles long) think of themselves as an endangered species. A recent newspaper headline said, "SHRIMPERS SPARED ANTI-TURTLE DEVICES." Even so, with the continuing wanton depletion of shrimp beds, they will undoubtedly have to find some other means of employment soon. They might, for instance, become part of that vast throng laboring in the *tourist industry.*

• • •

Tourism has become an industry as destructive as any other. You are no longer benign in traveling somewhere to look at the scenery. You never thought there was much gain in just looking anyway; you've always preferred to *use* the scenery in some manner. In your desire to get away from what you've got, you've caused there to be no place to get away *to.* You're just all bumpered up out there. Sewage and dumps have become prime indicators of America's lifestyle. In resort towns in New

England and the Adirondacks, measuring the flow into the sewage plants serves as a business barometer. Tourism is a growth industry. You believe in growth. *Controlled growth*, of course. Controlled exponential growth is what you'd really like to see. You certainly don't want to put a moratorium or a cap on anything. That's illegal, isn't it? Retro you're not. You don't want to go back or anything. Forward. Maybe ask directions later. Growth is *desirable* as well as being *inevitable*. Growth is the one thing you seem to be powerless before, so you try to be realistic about it. Growth—it's weird—it's like cancer or something.

As a tourist you have long ago discovered your national parks and are quickly *overburdening* them. All that spare land, and it belongs to you! It's exotic land too, not looking like all the stuff around it that looks like everything else. You want to take advantage of this land, of course, and use it in every way you can. Thus the managers—or *stewards*, as they like to be called—have developed *wise* and *multiple-use* plans, keeping in mind exploiters' interests (for they have their needs, too), as well as the desires of the backpackers. Thus mining, timbering, and ranching activities take place in the national forest, where the Forest Service maintains a system of logging roads eight times greater than the interstate highway system. Snowmobilers demand that their trails be *groomed*. The national parks are more of a public playground and are becoming increasingly Europeanized in their look and management. Lots of concessions and motels. Paths paved to accommodate strollers. You deserve a clean bed and a hot meal when you go into the wilderness. At least, your stewards think that you do. You keep your stewards busy. Not only must they cater to your multiple and conflicting desires, they have to manage your wildlife *resources*. They have managed wildfowl to such an extent that, the reasoning has become, if it weren't for hunters, ducks would disappear. Duck stamps and licensing fees support the whole rickety duck management system. Yes! If it weren't for the people who kill them, wild ducks wouldn't exist! Many a manager believes that better wildlife *protection* is provided when wildlife is allowed to be shot. Conservation commissions can only oversee hunting when hunting is allowed. But wild creatures are managed in other ways as well. Managers track and tape and tag and band. They relocate, restock, and reintroduce. They cull and control. It's hard to keep it straight. Protect or poison? Extirpate or just mostly eliminate? Sometimes even the stewards get mixed up.

• • •

This is the time of machines and models, hands-on management and master plans. Don't you ever wonder as you pass that billboard advertising another MASTER PLANNED COMMUNITY just what master they are actually talking about? Not the Big Master, certainly. Something brought to you by one of the tiny masters, of which there are many. But you like these tiny masters and have even come to expect and require them. In Florida they're well into building a ten-thousand-acre city in the Everglades. It's a *megaproject*, one of the largest ever in the state. Yes, they must have thought you wanted it. No, what you thought of as the Everglades, the park, is only a little bitty part of the Everglades. Developers have been gnawing at this irreplaceable, strange land for years. It's like they just *hate* this ancient sea of grass. Maybe you could ask them about this sometime. Every tree and bush and inch of sidewalk in the project has been planned, of course. Nevertheless, because the whole thing will take twenty-five years to complete, the plan is going to be constantly changed. You can understand this. The important thing is that there be a blueprint. You trust a blueprint. The tiny masters know what you like. You like a *secure landscape* and *access to services*. You like grass—that is, lawns. The ultimate lawn is the golf course, which you've been told has "some ecological value." You believe this! Not that it really matters—you just like to play golf. These golf courses require a lot of watering. So much that the more inspired of the masters have taken to watering them with effluent, *treated* effluent, but yours, from all the condos and villas built around the stocked artificial lakes you fancy.

I really don't want to think about sewage, you say, but it sounds like progress.

It is true that the masters are struggling with the problems of your incessant flushing. Cuisine is also one of their concerns. Great advances have been made in sorbets—sorbet intermezzos—in their clubs and fine restaurants. They know what you want. You want A HAVEN FROM THE ORDINARY WORLD. If you're a NATURE LOVER in the West, you want to live in a WILD ANIMAL HABITAT. If you're eastern and consider yourself more hip, you want to live in a new town—a brand-new reconstructed-from-scratch town— in a house of NINETEENTH-CENTURY DESIGN. But in these new towns the masters are building, getting around can be confusing. There is an abundance of curves and an infrequency of through streets. It's the new wilderness without any trees. You can get lost, even with all the "mental bread crumbs" the masters scatter about as visual landmarks—the windmill, the water views, the various groupings of landscape "material." You are lost, you know. But you trust a Realtor will show you the way. There are many more Realtors than tiny masters, and many of them have to make do with less than a loaf—that is, trying to sell stuff that's already been built in an environment already "enhanced" rather than something being planned—but they're everywhere, willing to show you the path. If Dante returned to Hell today, he'd probably be escorted down by a Realtor talking all the while about how it was just another level of Paradise.

When have you last watched a sunset? Do you remember where you were? With whom? At Loews Ventana Canyon Resort, the Grand Foyer will provide you with that opportunity through lighting that is computerized to diminish with the approaching sunset!

• • •

The tiny masters are willing to arrange Nature for you. They will compose it into a picture that you can look at at your leisure, when you're not doing work or something like that. Nature becomes scenery, a prop. At some golf courses in the Southwest, the saguaro cactuses are reported to be repaired with green paste when balls blast into their skin. The saguaro can attempt to heal themselves by growing over the balls, but this takes time, and the effect can be somewhat . . . baroque. It's better to get out the pastepot. Nature has become simply a visual form of entertainment, and it had better look snappy.

Listen, you say, *we've been at Ventana Canyon. It's in the desert, right? It's very, very nice, a world-class resort. A totally self-contained environment with everything that a person could possibly want, on more than a thousand acres in the middle of zip. It sprawls but nestles, like. And they've maintained the integrity of as much of the desert ecosystem as possible. Give them credit for that. Great restaurant, too. We had baby bay scallops there. Coming into the lobby there are these two big hand-carved coyotes, mutely howling. And that's the way we like them, mute. God, why do those things howl like that?*

Wildlife is a personal matter, you think. The attitude is up to you. You can prefer to see it dead or not dead. You might want to let it mosey about its business or blow it away. Wild things exist only if you have the graciousness to allow them to. Just outside Tucson, Arizona, there is a structure modeled after a French Foreign Legion outpost. It's the International Wildlife Museum, and it's full of dead animals. Three hundred species are there, at least a third of them—the rarest ones—killed and collected by one C. J. McElroy, who enjoyed doing it and now shares what's left with you. The museum claims to be educational because you can watch a taxidermist at work or touch a lion's tooth. You can get real close to these dead animals, closer than you can in a zoo. Some of you prefer zoos, however, which are becoming bigger, better, and bioclimatic. New-age zoo designers want the animals to *flow right out into your space.* In Dallas there's a Wilds of Africa exhibit; in San Diego there's a simulated rain forest, where you can thread your way "down the side of a lush canyon, the air filled with a fine mist from 300 high-pressure nozzles . . ."; in New Orleans you've constructed a swamp, the real swamp not far away being on the verge of disappearing. Animals in these places are abstractions—wandering relics of their true selves, but that doesn't matter. Animal behavior in a zoo is nothing like natural behavior, but that doesn't matter, either. Zoos are pretty, contained, and accessible. These new habitats can contain one hundred different species—not more than one or two of each thing, of course—on seven acres, three, one. You don't want to see *too much* of anything, certainly. An *example* will suffice. Sort of like a biological Crabtree & Evelyn basket selected with you in mind. You like things reduced, simplified. It's easier to take it all in, park it in your mind. You like things inside better than outside anyway. You are increasingly looking at and living in proxy environments created by substitution and simulation. *Resource economists* are a wee branch in the tree of tiny masters, and one, Martin Krieger, wrote, "Artificial prairies and wildernesses have been created, and there is no reason to believe that these artificial environments need be unsatisfactory for those who experience them. . . . We will have to realize that the way in which we experience nature is conditioned by our society—which more and more is seen to be receptive to responsible intervention."

• • •

Fiddle, fiddle, fiddle. You support fiddling, as well as meddling. This is how you learn. Though it's quite apparent that the environment has been grossly polluted and the natural world abused and defiled, you seem to prefer to continue pondering effects rather than preventing causes. You want proof, you insist on proof. A Dr. Lave from Carnegie-Mellon—and he's an expert, an economist and an environmental expert—says that scientists will have to prove to you that you will suffer if you don't become less of a "throw-away society." *If you really want me to give up my car or my air conditioner, you'd better prove to me first that the earth would otherwise be uninhabitable,* Dr. Lave says. Me is you, I presume, whereas you refers to them. You as in me—that is, me, me, me—certainly strike a hard bargain. Uninhabitable the world has to get before you rein in your requirements. You're a consumer after all, the consumer upon whom so much attention is lavished, the ultimate user of a commodity that has become, these days, everything. To try to appease your appetite for proof, for example, scientists have been leasing for experimentation forty-six pristine lakes in Canada.

They don't want to *keep* them, they just want to *borrow* them.

They've been intentionally contaminating many of the lakes with a variety of pollutants dribbled into the propeller wash of research boats. It's one of the *boldest experiments in lake ecology ever conducted.* They've turned these remote lakes into huge *real-world test tubes.* They've been doing this since 1976! And what they've found so far in these preliminary studies is that pollutants are really destructive. The lakes get gross. Life in them ceases. It took about eight years to make this happen in one of them, everything carefully measured and controlled all the while. Now the scientists are slowly reversing the process. But it will take hundreds of years for the lakes to recover. They think.

• • •

Remember when you used to like rain, the sound of it, the feel of it, the way it made the plants and trees all glisten? We needed that rain, you would say. It looked pretty too, you thought, particularly in the movies. Now it rains and you go, Oh-oh. A nice walloping rain these days means *overtaxing our sewage treatment plants*. It means *untreated waste discharged directly into our waterways*. It means . . .

Okay. Okay.

Acid rain! And we all know what this is. Or most of us do. People of power in government and industry still don't seem to know what it is. Whatever it is, they say, they don't want to curb it, but they're willing to study it some more. Economists call air and water pollution "externalities" anyway. Oh, acid rain. You do get so sick of hearing about it. The words have already become a white-noise kind of thing. But you think in terms of *mitigating* it maybe. As for the greenhouse effect, you think in terms of *countering* that. One way that's been discussed is the planting of new forests, not for the sake of the forests alone, oh my heavens, no. Not for the sake of majesty and mystery or of Thumper and Bambi, are you kidding me, but because, as every schoolchild knows, trees absorb carbon dioxide. They just soak it up and store it. They just love it. So this is the plan: you can plant millions of acres of trees, and you go on doing pretty much whatever you're doing—driving around, using staggering amounts of energy, keeping those power plants fired to the max. Isn't Nature remarkable? So willing to serve? You wouldn't think it had anything more to offer, but it seems it does. Of course, these "forests" wouldn't exactly be forests. They would be more like trees. *Managed trees.* The Forest Service, which now manages our forests by cutting them down, might be called upon to evolve in its thinking and allow these trees to grow. They would probably be patented trees after a time. Fast-growing, uniform, genetically created toxin-eating *machines*. They would be *new-age* trees, because the problem with planting the old-fashioned variety to combat the greenhouse effect, which is caused by pollution, is that they're already dying from it. All along the crest of the Appalachians from Maine to Georgia, forests struggle to survive in a toxic soup of poisons. They can't *help* us if we've killed them, now can they?

• • •

All right, you say, *wow, lighten up, will you? Relax. Tell about yourself.*

Well, I say, I live in Florida . . .

Oh my God, you say. Florida! Florida is a joke! How do you expect us to take you seriously if you still live there! Florida is crazy, it's pink concrete. It's paved, it's over. And a little girl just got eaten by an alligator down there. It came out of some swamp next to a subdivision and carried her off. That set your Endangered Species Act back fifty years, you can bet.

I . . .

Listen, we don't want to hear any more about Florida. We don't want to hear about Phoenix or California's Central Valley. If our wetlands—our vanishing wetlands—are mentioned one more time, we'll scream. And the talk about condors and grizzlies and wolves is becoming too de trop. We had just managed to get whales out of our minds. Now there are butterflies, frogs even that you want us to worry about. And those manatees. Don't they know what a boat propeller can do to them by now? They're not too smart. And those last condors are pathetic. Can't we just get this over with?

Aristotle said that all living beings are ensouled and strive to participate in eternity.

Oh, I just bet he said that, you say. *That doesn't sound like Aristotle. He was a humanist. We're all humanists here. This is the age of humanism. Militant humanism. And it has been for a long time.*

• • •

You are driving with a stranger in the car, and it is the stranger who is behind the wheel. In the backseat are your pals for many years now—DO WHAT YOU LIKE and his swilling sidekick, WHY NOT. A deer, or some emblematic animal—something from that myriad natural world you've come from that you now treat with such indifference and scorn—steps from the dimming woods and tentatively upon the highway. The stranger does not decelerate or brake, not yet, maybe not at all. The feeling is that whatever it is will *get out of the way.* Oh, it's a fine car you've got, a fine machine, and oddly you don't mind the stranger driving it, because in a way, everything has gotten too complicated, way, way out of your control. You've given the wheel to the masters, the managers, the comptrollers. Something is wrong, *maybe*, you feel a little sick, *actually*, but the car is luxurious and fast and you're *moving*, which is the most important thing by far.

Why make a fuss when you're so comfortable? Don't make a fuss, make a baby. Go out and get something to eat, build something. Make *another* baby. Babies are cute. Babies show you have faith in the future. Although faith is perhaps too strong a word. They're everywhere these days; in all the crowds and traffic jams, there are the babies too. You don't seem to associate them with the problems of population increase. They're just babies! And you've come to believe in them again. They're a lot more tangible than the afterlife, which, of course, you haven't believed in in ages. At least not for yourself. The afterlife now belongs to plastics and poisons. Yes, plastics and poisons will have a far more extensive afterlife than you, that's known. A disposable diaper, for example, which is all plastic and wood pulp, will take around four centuries to degrade. But you like disposables—so easy to use and toss—and now that marketing is urging you not to rush the potty training by making diapers for four-year-olds available and socially acceptable, there will be more and more dumped diapers around, each taking, like most plastics, centuries and centuries to deteriorate. In the sea, many marine animals die from ingesting or being entangled in discarded plastic. In the dumps, plastic squats on more than 25 percent of dump space. But your heart is disposed toward plastic. Someone, no doubt the plastics industry, told you it was convenient. This same industry avidly promotes recycling in an attempt to get the critics of their nefarious, multifarious products off their backs. That should make you feel better, because *recycling* has become an honorable word, no longer merely the hobby of Volvo owners. The fact is that people in plastics are born obscurants. Recycling won't solve the plastic glut, only reduction of production will, and the plastics industry isn't looking into that, you can be sure. Waste is not just the stuff you throw away, of course, it's also the stuff you use to excess. With the exception of *hazardous waste*, which you do worry about from time to time, it's even thought that you have a declining sense of emergency about the problem. Builders are building bigger houses because you want bigger. You're trading up. Utility companies are beginning to worry about your constantly rising consumption. Utility companies!

You haven't entered a new age at all but one of upscale nihilism, deluxe nihilism.

With each election there is the possibility that the environment will become a political issue. But it never does. You don't want it to be, preferring instead to continue in your politics of subsidizing and advancing avarice. The issues are the same as always—jobs, defense, the economy the economy the economy, maintaining the standard of living in this greedy, selfish, expansionistic, industrialized society.

You're getting a little shrill here, you say.

You're pretty well off. And you expect to become even better off. You do. What does this mean? More software, more scampi, more square footage, more communication towers to keep you in touch and amused and informed? You want to count birds? Go to the bases of communication towers being built all across the country. Three million migratory songbirds perish each year by slamming into towers and their attendant guy wires. The building of thousands of new digital television towers one thousand feet and taller is being expedited by the FCC, which proposes to preempt all local and state environmental laws. You have created an ecological crisis. The earth is infinitely variable and alive, and you are moderating it, simplifying it, killing it. It seems safer this way. But you are not safe. You want to find wholeness and happiness in a land increasingly damaged and betrayed, and you never will. More than material matters. You must change your ways.

What is this? Sinners in the Hands of an Angry God?

The ecological crisis cannot be resolved by politics. It cannot be resolved by science or technology. It is a crisis caused by culture and character, and a deep change in personal consciousness is needed. Your fundamental attitudes toward the earth have become twisted. You have made only brutal contact with Nature; you cannot comprehend its grace. You must change. Have few desires and simple pleasures. Honor nonhuman life. Control yourself, become more authentic. Live lightly upon the earth and treat it with respect. Redefine the word *progress* and dismiss the managers and masters. Grow inwardly and with knowledge become truly wiser. Think differently, behave differently. For this is essentially a moral issue we face, and moral decisions must be made.

A moral issue! *Okay, this discussion is now over. A moral issue . . . And who's this we now? Who are you, is what I'd like to know. You're not me, anyway. I admit someone's to blame and something should be done. But I've got to go. It's getting late. Take care of yourself.*

Key Lois

In the Florida Keys, just off Cudjoe in the Atlantic, is Loggerhead Key, which looks a little odd, as well it might since its only occupants are monkeys. One half of the key is gray, as though in shadow; the other half is a healthy, mangrove-wooded green. A narrow tidal river separates the two, a river that, it is said by those who put the monkeys on the island in the first place, they prefer not to cross. The monkeys, about sixteen hundred of them, inhabit the gray half, which they have denuded with their climbing, scrambling, swinging, and clinging. They are seventy-pound, dog-faced rhesus monkeys, bred and raised for laboratory research. As with the cattle one sees grazing in green pastures, they undoubtedly lead a happy life until they don't. There is no natural food for them on the island and no fresh water, so someone arrives daily by boat with water and hundreds of pounds of Purina Monkey Chow. At other times a boat will arrive, collect some of the monkeys to fill an order from a laboratory, and take them away. The monkeys are purportedly the biggest, blondest, and healthiest in the world.

The original colony of one hundred was captured in India in 1972. The descendants are now a multimillion-dollar business for the owners, one of whom wanted to rename Loggerhead Key, *Key Lois*, as a tribute to his wife, although why Lois would want an island full of howling monkeys named after her is not known. The state, not keen on having places renamed willy-nilly for uxorious reasons, refused the request. The owner got his way nonetheless with the acronym Laboratory Observing Island Simians, and Loggerhead Key became Key Lois.

In 1997 the state began formal proceedings to have the monkeys removed, contending that they were destroying mangroves and polluting the waters with their waste. "It's inconsistent," a community affairs advocate argued, "to continue allowing the monkeys to pollute the island while the state requires treatment of human waste." Like, fair is fair. One monkey had the misfortune to escape to nearby Little Crane Key, which is a frigate bird nesting area, and was shot by a US Fish and Wildlife officer. The suppliers of the animals to the labs had always claimed that the monkeys could not escape the island because they "don't like salt water," but this poor voyaging primate was identified by his tattoo.

Safariland

THE DESIRED ILLUSION HERE IS . . . AFRICA. THE REALIZATION that it is Africa contributes to only part of the effect. For it is not enough anymore that it is Africa, which is, in great part, a sad landscape, scorched, dispirited, full of people and cattle. Cattle and people are just cattle and people, after all. It gets harder and harder to muster up much enthusiasm for them. The Africa of the desired illusion is this: You have entered a portion of the earth that wild animals have retained possession of. The illusion here is that wild animals exist.

This perception is key to the safari package. People go on safari to gaze at the animals. There are still those who go on safari to kill animals or, as they sometimes like to say, "take them out." (*I was elated . . . it took all five of us to lift the king of beasts into the back of the Toyota. . . .*) But these groups are to be avoided. These groups have their own "concessions." Much of the planning that goes into the safari is choosing the best "concession" and avoiding those traveling in other safari groups. The mystique of the safari is its exclusiveness. This is a *once in a lifetime experience.* You don't want to have to share it with a mob.

Much of Africa no longer provides the desired safari experience. West Africa is cities, people. You can travel thousands of miles without seeing any animals at all. Some baboons maybe.

To many Africans, wild animals are a thing of the past, standing in the way of progress—progress perceived to be cows and goats. Grasslands that for thousands of years have supported hundreds of thousands of migrating beasts have turned into desert under the implacable browsing of livestock. Then there are the wars—Zaire, Angola, Uganda, everywhere surly, scattered violence—with soldiers gunning down animals whose hides or horns or tusks can be marketed to buy more machine guns. In South Africa you can safari in Kruger National Park, but it isn't quite . . . "correct." Of course, East Africa hasn't been what it used to be for years. Kenya has been written off by many travelers as being just too silly, preposterous even, its remaining game exhausted, it's easy to imagine, from being looked at so much. For one reason or another, a great deal of Africa no longer lends itself to the "ultimate" safari adventure. A "new" area of Africa had to be found for the safariphile, had to become, in fact, the Africa for this type of tourist, and in this regard, Botswana agreed to be found.

Not all of Botswana, of course, just a small part, compactly arranged in the north and including specifically the Okavango Delta, the Moremi Wildlife Reserve, and Chobe National Park. These areas can nicely supply the aura of the Africa of old. The government of Botswana knows what it wants in the way of tourist traffic through these areas—high cost and low density. It wants international tourists with money, and not too many of them, and it wants them touring where they can marvel at the fish eagle of the Okavango (an impressive bird), the big cats of Moremi (for what's a safari without lions?), and the elephants of Chobe (67,000—the largest concentration in

Africa). Why would the safariphile want to go anywhere else in Botswana anyway? Much of the west and the south and the vast central wilderness of the Kalahari Desert is without roads. Little is there at all but the Fence. The veterinary cordon fence, five feet high, made of high-tensile-strength steel wire, 1,875 miles of it. The indestructible, intractable, infamous Fence. The government started erecting the Fence in 1954 to segregate cattle from wild herds and to protect the former from the possible transmission of hoof-and-mouth disease, and has been extending it ever since. Hundreds of thousands of wild animals have died against it in their futile trek toward water in time of drought. The Fence runs everywhere, and where the Fence runs, the wild animals do not. The zebra is extinct now in all of Botswana except the north. The buffalo, too. There are no real free-roaming herds in Botswana anymore; the great migrations are over. The business of raising cattle has had disastrous ecological consequences in Africa, nowhere perhaps more acutely than in Botswana. Once occurring "widely and continuously," as the guidebooks would say, the wild animals are now broken up into numerous isolated populations. They are "scattered . . . restricted . . . formally present . . . marginally present . . . fragmented . . . markedly discontinuous . . . isolated . . . increasingly isolated." They are "absent in all of Botswana except the north."

• • •

A first-class camping safari takes about two weeks and begins and ends in Victoria Falls, Zimbabwe. Groups move in shifts from east to west or west to east during most of the year, but the dry season, from June through October, is when the congregation of animals is greatest. Each group is discreetly accompanied by a small staff that sets up the tents, digs the latrines, bakes the bread, gathers the wood, and heats the water for the bush showers. All the clients have to do is look.

In the beginning, the clients look at one another. None appears particularly charming or discerning to the other. This is one of those swift, initial, gloomy insights that is not altered by the passage of time. Everyone drinks for a while at the outside bar. Then they go to the inside bar. This bar is called Totoba, which means "to totter." On the wall hang old prints showing a variety of animals staggering about, not from the effects of booze, of course, but from having just been shot. Elephants about to crash, zebras in their death gallop, buffalo toppling over Victoria Falls, lions and tigers all at their moment of collapse. The bar is full of guides, mostly from the Zambezi River. They're having a party, and at midnight they'll all take their clothes off. This is what they always do. A few minutes after midnight, they'll put their clothes back on again. This bar is not full of "colorful bush characters" like the bar at Riley's Hotel in Maun, farther along the safari circuit, but the people here try. "Photographing animals instead of shooting them," one of these madcap guides confides, "is like flirting instead of fucking."

A bit bemused, the clients go back to their hotel rooms, turning their keys in a dozen locks. The key rings are made from the nut of the ilala palm. It looks just like ivory; it's called vegetable ivory. Tonight is just a room, but tomorrow they will be in the bush, "under the canvas," on safari. Tomorrow their adventure will begin.

In the morning the groups split and meet their guides. Group B's guide's name is Chunk, say. All guides can't be called Ian or Gavin or Colin. Group A goes off with their Colin. They will drive in a big open Land Rover to Chobe National Park and camp a few nights in the Serondela area on the beautiful Chobe River. Then they'll drive to Savuti. They don't yet know how depressing this will be. The guide knows, that is, but they don't. It will be a long, hot, and dusty drive to Savuti. They will constantly be fishing around in the cooler for cans of Coke or beer or apple juice or club soda. They will picnic along the way beside a somewhat soiled-looking wetland. They will stop at a village and take pictures of huts constructed out of mud, and Coke, beer, apple juice, and club soda cans. ("Wonderful insulating properties," Chunk will say when Group B sees this phenomenon.) Savuti used to be synonymous with the waters of its channel and its marsh. It was a natural concentration point during the dry season for herds of animals coming from the north, south, and east. Thousands of zebras and wildebeests migrated here, their abundance supporting large numbers of predators—lions, leopards, wild dogs. It was a fabulous place. But for ten years the channel has been dry, and now the land is sere and ragged. Only a few old bull elephants and some thin, listless lions still come down every twilight to a single muddy hollow into which a bit of water is pumped by the parks department.

Group A can't wait to hasten out of the starved Savuti. The toilets weren't very nice, and the food wasn't very good, either. Noodles and ham and corn fritters and fried eggs for dinner. Plus some kind of pink, peculiar, cold dessert. What was that? The cooks must be crazy. Group A is eager to get on the tiny, tippy plane that will get them out of there—a short flight over plains and pans to an airstrip in the Moremi Wildlife Reserve. The Moremi was set aside as a sanctuary by the local tribes in the 1960s when it became clear that hunting was wiping out the animals. There are many small camps scattered here, run by many different outfitters. This camp is called Mombo, and it is a classic bush camp, the folding chairs of the sunset hour facing the creamy, golden veld. They will like Moremi a lot. They will see cheetahs, and cheetah cubs. They will see giraffes. They will go on night drives, which were against the rules in Chobe, through the forests and savannas of Chiefs Island, and they will see animal eyes, thousands of animal eyes shining. They don't want to leave Moremi. This is the way they had dreamed it would be. They will have heard the groan and cough of lions at night as they lay on their cots, just like they were told they would.

But too soon they must leave. It's part of the plan, the experience. They must go farther into the delta, into the subtle, silent country of the Okavango, a great watery wilderness tangle of lagoons, twisting channels, and thousands of islands, some sandy and fringed with phoenix palms, others cool and dark with ebony trees and fig stands and forests of baobab. They will fly over this trackless expanse—hard to grasp but beautiful, they'll agree, original. Hazy and green, white flocks of egrets and storks wheeling below the droning airplane, gray droves of elephants resembling thickets of trees, red antelope dashing through silver water. They will land at a village called Jedibe, which they will be told means "ostrich shit" in the local language—difficult to believe, but they won't dispute it—and they will be transported from there in dugout canoes called mokoros to their camp in the delta's panhandle. They will also be told to call the native guides, the polers of the mokoros, Trust and Pilot and Clever. The name of their cook at camp is Warm, for heaven's sake. This is silly, but Group A will accept this and smile distractedly. They'll have a lot to remember the next few days as they are poled interminably about in the mokoros. There are 550 bird species in Botswana—so many that they even have their own book, by Kenneth Newman, called, not unreasonably, *Birds of Botswana*. This book is referred to over and over again in the many hours spent in the mokoros. Floating the Okavango is as peaceful as a coma must be, and it is only the scream of the birdwatchers that disturbs the limpid calm. "Malachite kingfisher!" the initiated will call out. "Lesser jacana! Red-shouldered widow! Paradise flycatcher! Wattled plover! Sharp-billed honeyguide," a perhaps over-initiated birder will cry, "*Birds of Botswana*, page two hundred and ten!"

They are poled about at lily-pad level through towering papyrus mazes. There is little aquatic life below the papyrus, for light can't penetrate the packed, rotting roots, and the water is acidic and low in oxygen. Any creature larger than a rat would sink through the floating mats. In fact, the papyrus swamp is preferred habitat for very few species. The birds have to fly elsewhere in the delta to feed. There are, however, two medium-size antelope that live on the delta's tiny islands, and it will take Group A days before they can pronounce their names with any confidence. These are the lechwe and the sitatunga. Even when these names are mastered, Group A will have great difficulty in telling the creatures apart because though they are very different in appearance, they are seldom seen. They are heard, sometimes, vanishing, and the polers confuse things further by sometimes calling out "Lechwe!" and sometimes "Sitatunga!" It wears them out in Group A trying to be the first to identify something or to even remember what someone else has identified. Even at breakfast there's no peace, for the "Heuglin's robin!" and the "grey lourie!" are chattering away. One bird calls, "Go away go away go away," and the other says, "Don't you do it don't you do it don't you do it," and there will always be someone who won't know which bird says what. They'll be glad to bid farewell to the mokoros and get back on the bush plane again, first to Maun, a gritty, messy abattoir town, and then back to Victoria Falls.

They're off safari now, their concerns elsewhere. Should they go to the Falls Crafts Village and watch Makishi tribes-men stilt-and-pole dancing? They've heard it's breathtaking. Should they have their fortune told by the *n'anga* in the zebra-skin skirt? Should they dare to eat crocodile thermidor?

• • •

All the while, Group B has been having the same experience, but in reverse. They begin with the mokoros, which have about them, actually, very much the feeling of eternity. They see the sun rise from the mokoros, and they see the sun set from the mokoros. Sometime in between they stand around the enor-mous termite mound in the middle of camp. There are four hundred species of termites in Africa, and although they don't have their own book, they are responsible for making many of the islands in the delta. The fungus-growing termite makes the largest mounds—big ragged castles that are as startling in the landscape as giraffes or baobab trees. The termites are constantly in the process of building their "termitaria," growing their elaborate fungus gardens in arched chambers, and lovingly tending their huge queen and her consort, who live in a royal cell at the center of the mound. This is explained somehow by Chunk, but the group finds it difficult to appreciate the magnitude of the process. Instead they take pictures of the mound from every possible angle. They take pictures of the mokutshumo tree, the motsaudi tree, the mokoba tree, the motshaba tree. At night they sit under the sausage tree and shine flashlights at the big dog-faced bats that come to drink nectar from the dishy red flowers. Later, they retire to their tents and spray the air with cans of Doom Super and Peaceful Sleep so that they will not have to hear, in the words of the old Boer explorer Laurens van der Post, "the mosquitoes singing their wild pagan hymn."

Actually, they don't hear mosquitoes at all, and this may have less to do with their personal cans of Doom Super than with the accumulation of chemicals that the government has been dumping on the Okavango for years in its effort to eliminate the tsetse fly. Chemicals are considered a great advance over earlier attempts at eradication, which included cutting down the trees to deprive the fly of cover and killing buffalo and kudu to deprive it of food. A chemical "cocktail" of deltamethrin and endosulfan is being used these days, sprayed from planes in a fine, low-volume mist. Even now, as Group B lies under the canvas in the delta wilderness in what they've been told is the last of old Africa, a plane flies over, misting. Not that they miss the singing of the wild pagan hymn. In fact, if they had to contend with mosquitoes as they were being poled about in the mokoros, they believe they would go out of their minds. They don't even want to take pictures of the mokoros anymore, they want to do something else, they want to see the Bushmen. They've been told they have the option of flying to the Tsodilo Hills in the bush country on the northwestern edge of the delta to see the descendants of the original inhabitants of Africa. They want to see them, the Bushmen, and those rock paintings that are twenty-five hundred years old. But Chunk says that it's not an interesting cultural experience anymore, that it's disappointing, that the Bushmen wear Levi's and Reeboks. "Modern Bushmen are a waste of time," Chunk says. So they don't see the Bushmen.

Chunk suggests that the day be spent instead searching for a Pel's fishing owl or a pangolin, a rare anteater with a tiny head whose large brown scales are composed of gluey-looking hair. The creature is rare because it is so "highly regarded" by the Africans, who believe that its hairy plates bring luck in love. A search for something that there is very little likelihood of finding elevates the purpose of any journey, and even this diffident group is beginning to suffer from a certain absence of purpose. They will not see the owl or the pangolin, but when they finally put the mokoros behind them and fly over the delta (agreeing that the overview is the best view) to Moremi, they will clamber into 4-wheel-drive vehicles once more, and see lions and wild dogs and even a honey badger, which is supposed to have a snarl more hideous than any other sound in

Africa. They won't hear it snarl, unfortunately, but they will see lions mating. They will park very close to these mating lions, as a matter of fact, and luckily the lions won't be doing it in the shade, but right out in the sun. The pictures will come out even better. They will see lions with a warthog kill, though what they'd really like is to witness a kill. A lion slamming a zebra up against the Land Rover or tearing out an impala's throat right before their eyes. That's what some people observe on trips like this, they've heard. So, it's possible they could witness such an ultimately African event. But they won't.

Still, they have a real sense that they're on safari now. A native guide accompanies Chunk in Moremi—his name is Ishmael. He's observant, he's an excellent tracker, he knows all the signs. He would never confuse a giraffe's feces with a kudu's, as many people would. He can tell the difference between a zebra's and a warthog's, and that's not easy either. Group B goes for walks with "Ish," and he points out all these things. They see porcupine quills without the porcupine, a neat bundle of them, and Ish says, "Lion." They see bone-white dung, and Ish says, "Hyena." If the dung is long, it means the animal hasn't had water for a while. If it's black, it means the animal's consumed a lot of blood. It's almost relaxing, learning all this esoteric stuff—they've heard that safaris are good for relieving stress. They get up each morning half an hour before first light to be driven around for five hours, then after brunch and a nap and a tea, they are driven around again. Sometimes, just when they're getting ready for their naps, Chunk will say, "Ish has seen some vultures, guys, let's go!" and they will pile into the truck to see something it's all over for, whatever it was. "Mongoose," Ish says. "Steenbok." Chunk drives right up to everything—aardwolves, monitor lizards, trees full of bee eaters. The only thing he doesn't drive up to is another guided safari vehicle, particularly if it's stopped. It's the rule: if one group is looking at something, the other group can't look at the same time.

Eventually, Group B has to leave Moremi, as all groups must, and approach the Savuti camp. There is the sound of tent flaps being zipped and unzipped and the sound of film rewinding. They've found that gin and apricot juice doesn't make a bad drink—the quinine cans have too much the odor of the spoiling sausage they nestle beside in the cooler—and some terrible jokes are told around the campfire. There are elephant-dick jokes and ostrich-dick jokes and even jokes comparing mice and pandas. Chunk doesn't tell any jokes, but he tells a few stories concerning his charisma, his adventures in other faraway places, and his aplomb with wild animals. Someone sings. The clients seem to have found themselves, and what they've found is that they're a rather shrieky lot. Giggling and chortling and laughing and chuckling, they endure the days in Savuti's barren marsh and, after a brief, rollicking journey, arrive at the Chobe River. They go to the chic Chobe Game Lodge, where Elizabeth Taylor got married, not this last time but the other time, when she married Richard Burton the second time. They take a sunset river cruise. They see the animals come down to drink in long, eager lines. They see silent herds backlit by a big, cloud-streaked, solemn sun. Back at camp they find that baboons have broken into the tents. They've taken candy and prescription drugs. One woman, recently retired from the foreign service, is particularly upset. The baboons have stolen her nuts and scattered her cosmetics about. "I feel so violated," she says.

But this is Africa, this really feels like it could be Africa at last. The air is dazzling, and in the morning the hippos are honking in the river and the tawny fields fairly ripple with life. The trees are full of nests—weaver nests, vulture nests, eagle nests, the extravagant domed nests of the drab little hammerkop. The members of Group B set off on a game drive. They see trumpeter hornbills, a pair of them. Not at all common. They see secretary birds, which they've previously seen only in *Birds of Botswana*. They see magnificent sable antelope, with their sweeping horns. They are tearing around in their Land Rover, happy as can be, filled with Africa, giddy with it. Springbok, steenbok, grysbok! Kudu, eland, who can tell the difference? Rollers! Hoopoes! they call out. They are claiming and naming them all, chattering away, glossing and glassing everything, and then they come upon the elephants, and they fall silent.

The elephants are fantastic. They're mysterious, they have dignity. Over and over throughout the day, they see elephants. Families with youngsters. Baby calves that can still walk under their mother's stomach. Young bulls. Groups, herds, hundreds of them on the move, swaying, silent, swinging their trunks. When a column of them crosses the road, a guard turns out and faces the vehicle, forefoot swinging, great ears flapping, prepared to do what's possible to defend the family as it hurries past. The air is electric with elephant. The group takes pictures, shoots roll after roll, watching them. They could watch them for hours. They do watch them for hours, bathing in the river, feeding in the hills. The elephants enchant them.

Many things are known about the elephant. For example, it's known that its phalanges are embedded in a soft cushion of elastic fiber enclosed in cartilage, which serves as a shock absorber explaining the extraordinary silence of the animal's passage through the bush, and that its trunk is one of the most versatile instruments in nature. It's known that its gestation period is almost two years, after which a single young is produced. It's known that this young suckles for three to four years and that sexual maturity is not reached for ten years. It's known that elephants are led by a succession of mothers and daughters—that female elephants stay with their mothers all their lives. It's known that the leader in a family is the oldest cow, whose great experience is the herd's knowledge that protects it from drought and famine and sometimes humans. It's known that aggressive elephant behavior is almost always bluffing. It's known that so many elephants have been killed in the last decade that few of the remnant herds are guided by matriarchs old enough to have generations of memory in their heads. It's known that there were two million African elephants in 1970 and about 600,000 elephants are left. It could be said that these animals, mostly young and frightened and with little access to the accumulated knowledge of the past, have lost Africa. That in their heads, their Africa has vanished.

It's known that it's never been difficult to kill an elephant.

In his book *The Last Place on Earth*, Harold Hayes tells of how a group of elephants broke into a shed where the ears and feet of elephants destroyed in a "cropping" procedure were being stored before being made into handbags and umbrella stands, removed the remains, and buried them.

There are many stories like this—old, troubling stories. Elephants have a sense of propriety. They know grief and indignation. They are intelligent, they are telepathic, they can communicate across many miles through infrasonic sound. They are highly protective of one another and care for their young with exceptional tenderness. They know what death is. It worries and mystifies and upsets them. Or, as Peter Matthiessen says in *African Silences*, they "are increasingly being credited with the apprehension of death."

When elephants are "cropped" (as though "cutting back" their number was as bloodless and sensible as pruning a tree, strictly an agricultural procedure), whole families—cows, calves, and juveniles—are removed en masse by "highly efficient teams." It usually takes a few minutes. The croppers employing their cropping tool, the machine gun, confine themselves to selected families in order to prevent panic from spreading throughout the entire elephant population.

"That's sad," someone says, "really sad."

But as a character in Robert Ruark's old, unclassic novel *Uhuru* remarks, "If you were there to kill, some unpleasantness was necessary."

Tourism is an avenue for political greed and limitless exploitation in Africa, it's true, but there are fences and farmers too. There's tsetse control and game control, dams and poachers and government croppers. There are the field biologists who tag and track things, and volunteers—they don't want to be just tourists—who spend their adventure holiday tagging and tracking things for them. There are hunters who now pay to shoot rhinos with $7,000 tranquilizing darts so that conservationists can insert microchips into their horns to monitor their movements. It's chaotic, all these people trying to keep up their interests. It makes the tourist—the safariphile—seem almost innocent.

Some of the ladies in Group B have become quite taken with Chunk and tip him $100 at the end of the trip. The men tip him $50. The safari cost each around $3,000 plus airfare, but what an experience—hasn't it been an experience? They're all back in Victoria Falls now, and they decide to go to the bar for a farewell drink. Some hunters are there, and one of them is saying, "With buffalo the problem is not so much killing them as getting them to understand that they have undergone a change of status."

Another one is talking about hunting forest elephants somewhere in Africa the year before. He came across one and could have shot it easily, but he passed it up because he thought it was too small. Now he kind of wishes he had taken it. The days of the big tuskers are over. He decries their indiscriminate killing by poachers. "My six-year-old son, Joe, will never get the chance that I passed up," he says.

Is this macho pathos? Is it the new thoughtfulness?

The safariphiles decide that they really hate this bar, that they're going to go into town. Should they go to the snake park? Should they buy a crocodile belt? Do they really dare to have the future told? Some of them, on reflection, think they might have taken too many pictures. It's easy to take too many pictures of the elephants. Someone worries that the shutter blades in his camera look a little off. I might have lost that last roll, he says. I could have lost all those elephants. I'll send you some of mine, someone says; I know I took too many pictures of the elephants. But then they agree that it's not the same. If you come on safari and take pictures of elephants, you want the ones you've taken and not someone else's. But then someone says, Oh, pictures, they never come out the way you saw it anyway. And everyone agrees. They agree that Africa proved to be pretty accessible, actually. There wasn't quite the astonishing number of animals they expected, but there were a lot. In Africa, wild animals still provide the opportunity to be seen. They're still out there, in Africa, offering people the experience of their existence. Isn't that nice?

Wildebeest

WILDEBEEST ARE A BIG STRANGE-LOOKING ANTELOPE, A GNU. Their shoulders are higher and heavier than their rumps. They have stringy tails, and manes, and beards on their throats and are dark with streaks of silver. They have horns and gallop like horses and have big Cubistic heads with mismatched features. They're fabulously grotesque looking and they travel—they used to travel—in great migratory herds across Africa. In Botswana in the 1970s they died by the tens of thousands during the drought, and their populations have never recovered. It was not that there was no water. It was that the wildebeest and many other wild animals were prevented from reaching it by a fence. A cattle fence that angles and runs for hundreds of miles. A hundred thousand wildebeest which had been spread out across a vast wilderness were forced by the fence to take the same migratory route to water that they never would reach. Almost four hundred miles of river and lake shores that had once been available to them in droughts had been reduced to a few miles by the fence. The wildebeest plodded along the fence, dying all along the way until the fence turned away from the water they had been smelling for days, joining another fence and forming a corner in which most of the survivors collapsed. Their heads hanging, they tottered and fell, their eyes plucked out by vultures while they still lived, their ears and testicles chewed off by scavengers while their legs still moved, as though they were moving still toward the water. Suffer and suffer and die. Mythical biblical it almost seems. Well, that's what happened to the wildebeest, and is still happening to them. There are still droughts, and the wildebeest die against the fence, but not in the great numbers of the '70s because those great numbers no longer exist. When I think about Africa what I think is *wildebeest*—that wild, uncomprehending, incomprehensible thing that thirsts. And I think that when you're talking about darkness, the blackness of darkness, you're talking about wildebeest (not elephants, the monumental moral innocence of elephants, but wildebeest), for wildebeest are at the great empty heart of blackness, the heart of its nothingness, dying over and over again against the indifferent fence with the water just beyond it.

The Killing Game

DEATH AND SUFFERING ARE A BIG PART OF HUNTING. A BIG part. Not that you'd ever know it by hearing hunters talk. They tend to downplay the killing part. To kill is to put to death, extinguish, nullify, cancel, destroy. But from the hunter's point of view, it's just a tiny part of the experience. *The kill is the least important part of the hunt . . .* , they often say, or, *Killing involves only a split second of the innumerable hours we spend surrounded by and observing nature. . . .* For the animal, of course, the killing part is of considerably more importance. José Ortega y Gasset, in *Meditations on Hunting*, wrote, *Death is a sign of reality in hunting. One does not hunt in order to kill; on the contrary, one kills in order to have hunted.* This is the sort of intellectual blather that the "thinking" hunter holds dear. The conservation editor of *Field & Stream* once paraphrased this sentiment by saying, *We kill to hunt, and not the other way around*, thereby making it truly fatuous. A hunter in West Virginia, one Mr. Bill Neal, blazed through this philosophical fog by explaining why he blows the toes off treed raccoons so that they will fall down and be torn apart by his dogs: *That's the best part of it. It's not any fun just shooting them.*

Instead of monitoring animals—many animals in managed areas are tattooed and wear radio transmitters—wildlife managers should start hanging telemetry gear around hunters' necks to study their attitudes and record their conversations. It would be grisly listening, but it would tune out for good the suffering as sacrament and spiritual experience nonsense that some hunting apologists employ. *The unease with which the good hunter inflicts death is an unease not merely with his conscience but with affirming his animality in the midst of his struggles toward humanity and clarity*, Holmes Rolston III drones on in his book *Environmental Ethics*.

There is a formula to this in literature—someone the protagonist loves has just died, so he goes out and kills an animal. This makes him feel better. But it's kind of a sad feeling-better. He gets to relate to Death and Nature in this way. Somewhat. But not really. Death is still a mystery. Well, it's hard to explain. It's sort of a semireligious thing.... Killing and affirming, affirming and killing, it's just the cross the "good" hunter must bear. The bad hunter merely has to deal with postkill letdown.

Many are the hunter's specious arguments. Less semireligious but a long-standing favorite with them is the vegetarian approach (you eat meat, don't you?). If you say no, they feel they've got you—you're just an oddball attempting to impose your weird views on others. If you say yes, they accuse you of being hypocritical, of allowing your genial Safeway butcher to stand between you and reality. The fact is, the chief attraction of hunting is the pursuit and murder of animals—the meat-eating aspect of it is trivial. If the hunter chose to be *ethical* about it, he might cook his kill, but the meat of most animals is discarded. Dead bear can even be dangerous! A bear's heavy hide must be skinned at once to prevent meat spoilage. With effort, a hunter can make okay chili, *something to keep in mind*, a sports rag says, *if you take two skinny spring bears.*

As for subsistence hunting, please. . . . The subsistence line is such a cynical one. Your Pennsylvania sportsman might personally not want to shoot a whale, but he sure supports a Northwest Indian's whim to do so, because any breaching or weakening of laws prohibiting any type of hunting is in the interests of all hunters. The subsistence/cultural/traditional argument is as egregious as it is canny. The Makah Indian tribe of Washington won the right to kill the gray whale in 1999 despite the US ban on whaling, and since the whales in Neah Bay on the Olympic Peninsula were used to being approached by tourist boats, a successful hunt was pretty much assured. A young whale was easily killed with high-powered rifles and hauled ashore with the assistance of some commercial fishing boats, but the community hadn't a clue as to what to do with the two tons of meat that resulted, blubber not having been a part of the tribe's diet for over seventy years. After dancing on the dead whale's back for a while, everybody went home, leaving the meat to rot. Granted that there might be a few "good" hunters out there who conduct the kill as spiritual exercise and a few others who are atavistic enough to want to supplement their Chicken McNuggets with venison, most hunters hunt for the hell of it.

For hunters, hunting is fun. Recreation is play. Hunting is recreation. Hunters kill for play, for entertainment. They kill for the thrill of it, to make an animal "theirs." (The Gandhian doctrine of nonpossession has never been a big hit with hunters.) The animal becomes the property of the hunter by its death. Alive, the beast belongs only to itself. This is unacceptable to the hunter. *He's yours. . . . He's mine. . . . I decided to. . . . I decided not to. . . . I debated shooting it, then I decided to let it live. . . .* Hunters like beautiful creatures. A "beautiful" deer, elk, bear, cougar, bighorn sheep. A "beautiful" goose or mallard. Of course, they don't stay beautiful for long, particularly the birds. Many birds become rags in the air, shredded, blown to bits. *Keep shooting till they drop!* Hunters get a thrill out of seeing a plummeting bird, out of seeing it crumple and fall. *The big pheasant folded in classic fashion.* They get a kick out of "collecting" new species. *Why not add a unique harlequin duck to your collection?* Swan hunting is satisfying. *I let loose a three-inch Magnum. The large bird only flinched with my first shot and began to gain altitude. I frantically ejected the round, chambered another, and dropped the swan with my second shot. After retrieving the bird, I was amazed by its size. The swan's six-foot wingspan, huge body, and long neck made it an impressive trophy.* (The hunter might also have been amazed that he killed the wrong bird. When endangered trumpeter swans were moved by wildlife agencies from Wyoming's Yellowstone to other western states in the hope they would learn to migrate, they were shot by hunters who thought they were *just* tundra swans.) Hunters like big animals, trophy animals. A "trophy" usually means that the hunter doesn't deign to eat it. Maybe he skins it or mounts it. Maybe he takes a picture. *We took pictures, we took pictures.* Maybe he just looks at it for a while. The disposition of the "experience" is up to the hunter. He's entitled to do whatever he wishes with the damn thing. It's dead.

Hunters like categories they can tailor to their needs. There are the "good" animals—deer, elk, bear, moose—which are allowed to exist for the hunter's pleasure. Then there are the "bad" animals, the vermin, varmints, and "nuisance" animals, the rabbits and raccoons and coyotes and beavers and badgers, which are disencouraged to exist. The hunter can have fun killing them, but the pleasure is diminished because the animals aren't "magnificent."

Many people in South Dakota want to exterminate the red fox because it preys upon some of the ducks and pheasant they want to hunt and kill each year. They found that after they killed the wolves and coyotes, they had more foxes than they wanted. The ring-necked pheasant is South Dakota's state bird. No matter that it was imported from Asia specifically to be "harvested" for sport, it's South Dakota's state bird and they're proud of it. A group called Pheasants Unlimited gave some tips on how to hunt foxes: *Place a small amount of larvicide* [a grain fumigant] *on a rag and chuck it down the hole. . . . The first pup generally comes out in fifteen minutes. . . . Use a .22 to dispatch him. . . . Remove each pup shot from the hole. Following gassing, set traps for the old fox who will return later in the evening. . . .* Poisoning, shooting, trapping—they make up a sort of sportsman's triathalon.

In the hunting magazines, hunters freely admit the pleasure of killing to one another. *Undeniable pleasure radiated from her smile. The excitement of shooting the bear had Barb talking a mile a minute.* But in public, most hunters are becoming a little wary about raving on as to how much fun it is to kill things. Hunters have a tendency to call large animals by cute names— "bruins" and "muleys," "berry-fed blackies" and "handsome cusses" and "big guys," thereby implying a balanced jolly game of mutual satisfaction between hunter and the hunted—*Bam, bam, bam, I get to shoot you and you get to be dead.* More often, though, when dealing with the nonhunting public, a drier, businesslike tone is employed. Animals become a "resource" that must be "utilized." Hunting becomes "a legitimate use of the resource." Animals become a product like wool or lumber or a crop like fruit or corn that must be "collected" or "taken" or "harvested." Hunters love to use the word legitimate. (Oddly, Tolstoy referred to hunting as "evil legitimized.") *A legitimate use, a legitimate form of recreation, a legitimate escape, a legitimate pursuit.* It's a word they trust will slam the door on discourse. Hunters are increasingly relying upon their spokesmen and supporters, state and federal game managers and wildlife officials, to employ solemn bureaucratic language and toss around a lot of questionable statistics to assure the nonhunting public (93 percent!) that there's nothing to worry about. The pogrom is under control. The mass murder and manipulation of wild animals is just another business. Hunters are a tiny minority, and it's crucial to them that the millions of people who don't hunt not be awakened from their long sleep and become antihunting. Nonhunters are okay. Dweeby, probably, but okay. A hunter *can respect the rights* of a nonhunter. It's the "antis" he despises, those *misguided, emotional, not-in-possession-of-the-facts, uninformed zealots who don't understand nature. . . . Those dimestore ecologists cloaked in ignorance and spurred by emotion. . . . Those doggy-woggy types, who under the guise of being environmentalists and conservationists are working to deprive him of his precious right to kill.* (Sometimes it's just a *right*; sometimes it's a *God-given* right.) Antis can be scorned,

46

but nonhunters must be pacified, and this is where the number crunching of wildlife biologists and the scripts of *professional resource managers* come in. Leave it to the professionals. They know what numbers are the good numbers. Utah determined that there were six hundred sandhill cranes in the state, so permits were issued to shoot one hundred of them. Don't want to have too many sandhill cranes. California wildlife officials report "sufficient numbers" of mountain lions to "justify" renewed hunting, because the animal is not "rare." (It's always a dark day for hunters when an animal is adjudged *rare*. How can its numbers be "controlled" through hunting if it scarcely exists?) A citizens' referendum of 1994 prohibits the hunting of the mountain lion in perpetuity in that state, though hunting lobbies continue to finance ballot initiatives that would weaken if not nullify such protection. (To hear the wannabe lion hunters talk, the big cat's preferred habitat is directly outside the schoolyard gates.) With or without protection, lions are killed anyway, in California and all over the West annually by the government as a social courtesy to modern settlers carving suburbs in the wild and as a professional courtesy to ranchers. One dead calf usually results in four dead mountain lions. In Oregon, state biologists applied for a one million dollar grant to kill *fifty* mountain lions just to see if they could determine if fewer cougars would result in more deer and elk. Are cougars keeping huntable herd levels down or just eating animals that would have died anyway? Biologists want to know! In Montana, buffalo can be extirpated only by government shooters, though officials would like to reclassify them as big-game animals, which would qualify them to be shot by sportsmen. Montana has definite ideas concerning the number of buffalo that can be tolerated. Zero is the number. Montana is so annoyed that the creatures weren't eradicated a century ago. (Why didn't they thunder off into extinction? Then we could feel bad about them. . . .) In the winter of 1988, nearly six hundred buffalo wandered out of the north boundary of Yellowstone Park and into Montana, where they were immediately shot at point-blank range by lottery-winning hunters. It was easy. And it was obvious from a video taken on one of the blow-away-a-bison days that the hunters had a heck of a good time. Buffalo don't spook easily. In general, they just stand around in their massive fashion, making exceptionally easy targets. The video disturbed a lot of nonhunters—it just didn't seem *fair*—so Montana is laboring to formulate a "fair chase" hunt, which would entail getting those shaggy symbols of an unfettered West to *move*, maybe break into a trot. As well as wanting zero buffalo, Montana wants zero wolves. Large predators—including grizzlies, cougars, and wolves—are often the most "beautiful," the smartest and wildest animals of all. The gray wolf is both a supreme predator and an endangered species, and since the Supreme Court has affirmed that ranchers have no constitutional right to kill endangered predators— apparently some God-given rights are not constitutional ones—this makes the wolf a more or less lucky dog, though pretty much on paper. A small population of gray wolves has established itself in northwestern Montana, primarily in Glacier National Park, and a pilot "recovery" program continues in Arizona's White Mountains despite some disheartening initial results (of eleven wolves released, five were shot dead by persons unknown). It has long been a dream of conservationists to bring back the wolf to Yellowstone. A "reintroduction" plan is proceeding. But to please ranchers and hunters, part of the plan would involve immediately removing the wolf from the endangered-species list. Beyond the park's boundaries, the wolf could be hunted as a "game animal" or exterminated as a "pest." (Hunters kill to hunt, remember, except when they're hunting to kill.) The area of Yellowstone where the wolf would be restored is the same mountain and high-plateau country that is abandoned in winter by most animals, including the aforementioned luckless bison. Part of the plan, too, is compensation to ranchers if any of their far-ranging livestock is killed by a wolf. It's a real industry out there, apparently, killing and controlling and getting compensated for losing something under the Big Sky. Wolves gotta eat—a fact that disturbs hunters. Jack Atcheson, an outfitter in Butte, said, *Some wolves are fine if there is control. But there never will be control. The wolf-control plan provided by the Fish and Wildlife Service speaks only of protecting domestic livestock. There is no plan to protect wildlife. . . . There are no surplus deer or elk in Montana. . . . Their numbers are carefully managed. With uncontrolled wolf populations, a lot of people will have to give up hunting just to feed wolves. Will you give up your elk permit for a wolf?*

It won't be long before hunters start demanding compensation for animals they aren't able to shoot.

• • •

Hunters believe that wild animals exist only to satisfy their wish to kill them. And it's so easy to kill them! The weaponry available is staggering, and the equipment and gear limitless.

The demand for big boomers has never been greater than right now, Outdoor Life *crows, and the makers of rifles and cartridges are responding to the craze with a variety of light artillery that is virtually unprecedented in the history of sporting arms.* . . . Hunters use grossly overpowered shotguns and rifles and compound bows. They rely on all manner of vehicles that swiftly place them in otherwise inaccessible landscapes. . . . *He was interesting, the only moving living creature on that limitless white expanse. I slipped a cartridge into the barrel of my rifle and threw the safety off.* . . . They use snowmobiles to run down elk, and dogs to run down and tree cougars. It's easy to shoot an animal out of a tree. It's virtually impossible to miss a moose, a conspicuous animal of steady habits. . . . *I took a deep breath and pulled the trigger. The bull dropped. I looked at my watch: 8:22. The big guy was early. Mike started whooping and hollering and I joined him. I never realized how big a moose was until this one was on the ground. We took pictures.* . . . Hunters shoot animals when they're resting. . . . *Mike selected a deer settled down to a steady rest, and fired. The buck was his when he squeezed the trigger. John decided to take the other buck, which had jumped up to its feet. The deer hadn't seen us and was confused by the shot echoing about in the valley. John took careful aim, fired, and took the buck. The hunt was over.* . . . And they shoot them when they're eating. . . . *The bruin ambled up the stream, checking gravel bars and backwaters for fish. Finally he plopped down on the bank to eat. Quickly, I tiptoed into range.* . . . They use sex lures. . . . *The big buck raised its nose to the air, curled back its lips, and tested the scent of the doe's urine. I held my breath, fought back the shivers, and jerked off a shot. The 180-grain spire-point bullet caught the buck high on the back behind the shoulder and put it down. It didn't get up. . . .* They use walkie-talkies, binoculars, scopes. . . . *With my 308 Browning BLR, I steadied the 9X cross hairs on the front of the bear's massive shoulders and squeezed. The bear cartwheeled backward for fifty yards. . . . The second Federal Premium 165-grain bullet found its mark. Another shot anchored the bear for good. . . .* They bait deer with corn. They spread popcorn on golf courses for Canada geese, and they douse meat baits with fry grease and honey for bears. . . . *Make the baiting site redolent of inner-city doughnut shops.* They use blinds and tree stands and mobile stands. They go out in groups, in gangs, and employ "pushes" and "drives." So many methods are effective. So few rules apply. It's fun! . . . *We kept on repelling the swarms of birds as they came in looking for shelter from that big ocean wind, emptying our shell belts. . . .* A species can, in the vernacular, be *pressured by hunting* (which means that killing them has decimated them), but that just increases the fun, the *challenge.* There is practically no criticism of conduct within the ranks. . . . *It's mostly a matter of opinion and how hunters have been brought up to hunt. . . .* Although a recent editorial in *Ducks Unlimited* magazine did venture to primly suggest that *one should not fall victim to greed-induced stress through piggish competition with others.*

But hunters are piggy. They just can't seem to help it. They're overequipped . . . insatiable, malevolent, and vain. They maim and mutilate and despoil. And for the most part, they're inept. Grossly inept.

Camouflaged toilet paper is a must for the modern hunter, along with his Bronco and his beer. Too many hunters taking a dump in the woods with their roll of Charmin beside them were mistaken for white-tailed deer and shot. Hunters get excited. They'll shoot anything—the pallid ass of another sportsman or even themselves. A Long Island man died when his shotgun went off as he clubbed a wounded deer with the butt. Hunters get mad. They get restless and want to fire! They want to use those assault rifles and see foamy blood on the ferns. Wounded animals can travel for miles in fear and pain before they collapse. Countless gut-shot deer—*if you hear a sudden squashy thump, the animal has probably been hit in the abdomen*—are "lost" each year. "Poorly placed shots" are frequent, and injured animals are seldom tracked, because most hunters have never learned how to track. The majority of hunters will shoot at anything with four legs during deer season and anything with wings during duck season. Hunters try to nail running animals and distant birds. They become so overeager, so *aroused*, that they misidentify and misjudge, spraying their "game" with shots but failing to bring it down.

The fact is, hunters' lack of skill is a big, big problem. And nowhere is the problem worse than in the touted glamour recreation—bow hunting. These guys are elitists. They doll themselves up in camouflage, paint their faces black, and climb up into tree stands from which they attempt the penetration of deer, elk, and turkeys with modern, multiblade, broadhead arrows shot from sophisticated, easy-to-draw compound bows. This "primitive" way of hunting appeals to many, and even the nonhunter may feel that it's a "fairer" method, requiring more strength and skill, but bow hunting is the cruelest, most wanton form of wildlife disposal of all. Studies conducted by state fish and wildlife departments repeatedly show that bow hunters wound and fail to retrieve as many animals as they kill. An animal that flees, wounded by an arrow, will most assuredly die of the wound, but it will be days before he does. Even with a "good" hit, the time between the strike and death is exceedingly long. *The rule of thumb has long been that we should wait thirty to forty-five minutes on heart and lung hits, an hour or more on a suspected liver hit, eight to twelve hours on paunch hits, and that we should follow immediately on hindquarter and other muscle-only hits, to keep the wound open and bleeding,* is the advice in the magazine *Fins and Feathers.* What the hunter does as he hangs around waiting for his animal to finish with its terrified running and dying hasn't been studied—maybe he puts on more makeup, maybe he has a highball.

Wildlife agencies promote and encourage bow hunting by permitting earlier and longer seasons, even though they are well aware that, in their words, *crippling is a by-product of the sport,* making archers pretty sloppy for elitists. The broadhead arrow is a very inefficient killing tool. Bow hunters are trying to deal with this problem with the suggestion that they use poison pods. These poisoned arrows are illegal in all states except Mississippi (*Ah'm gonna get ma deer even if ah just nick the little bastard*), but they're widely used anyway. You wouldn't want that deer to suffer, would you?

The mystique of the efficacy and decency of the bow hunter is as much an illusion as the perception that a waterfowler is a refined and thoughtful fellow, a *romantic aesthete,* as the writer Vance Bourjaily put it, equipped with his faithful lab and a love for solitude and wild places. More sentimental drivel has been written about bird shooting than about any other type of hunting. It's a soul-wrenching pursuit, apparently, the execution of birds in flight. Ducks Unlimited—an organization that has managed to put a spin on the word *conservation* for years—works hard to project the idea that duck hunters are blue bloods and that duck stamps with their pretty pictures are responsible for saving all the saved puddles in North America. Sportsman's conservation is a contradiction in terms (We protect things now so that we can kill them later) and is broadly interpreted (Don't kill them all; just kill most of them). A hunter is a conservationist in the same way a farmer or a rancher is: He's not. Like the rancher who kills everything that's not stock on his (and the public's) land, and the farmer who scorns wildlife because "they don't pay their freight," the hunter uses nature by destroying its parts, mastering it by simplifying it through death.

The previously mentioned sports mag intellectual ("We kill to hunt and not the other way around") said that the "dedicated" waterfowler will shoot other game "of course," but they *do so much in the same spirit of the lyrics, that when we're not near the girl we love, we love the girl we're near.* (Duck hunters practice tough love.) The fact is, far from being a romantic aesthete, the waterfowler is the most avaricious of all hunters. . . . *That's when Scott suggested the friendly wager on who would take the most birds . . .* and the most resistant to minimum ecological decency. Millions of birds that managed to elude shotgun blasts were dying each year from ingesting the lead shot that rained down in the wetlands. Decade after decade, birds perished from feeding on spent lead, but hunters were "reluctant" to switch to steel. They worried that it would impair their shooting, and ammunition manufacturers said a changeover would be "expensive." State and federal officials had to weigh the poisoning against those considerations. It took forever, this weighing, but now steel-shot loads are required almost everywhere, having been judged "more than adequate" to bring down the birds. This is not to say, of course, that most duck hunters use steel shot everywhere. They're traditionalists and don't care for all the new, pesky rules. Oh, for the golden age of waterfowling, when a man could measure a good day's shooting by the pickup load. But those days are gone. Fall is a melancholy time, all right.

Spectacular abuses occur wherever geese congregate, quietly notes Shooting Sportsman, something that the more cultivated *Ducks Unlimited* would hesitate to admit. Waterfowl populations are plummeting, and waterfowl hunters are out of control. "Supervised" hunts are hardly distinguished from unsupervised ones. A biologist with the Department of the Interior who observed a hunt at Sand Lake in South Dakota said, *Hunters repeatedly shot over the line at incoming flights where there was no possible chance of retrieving. Time and time again I was shocked at the behavior of hunters. I heard them laugh at the plight of dazed cripples that stumbled about. I saw them striking the heads of retrieved cripples against fence posts.* In the South, wood ducks return to their roosts after sunset when shooting hours are closed. Hunters find this an excellent time to shoot them. Dennis Anderson, an outdoors writer, said, *Roost shooters just fire at the birds as fast as they can, trying to drop as many as they can. Then they grab what birds they can find. The birds they can't find in the dark, they leave behind.*

Carnage and waste are the rules in bird hunting, even during legal seasons and open hours. Thousands of wounded ducks and geese are not retrieved, left to rot in the marshes and fields.... *When I asked Wanda where hers had fallen, she wasn't sure.* Cripples, and many cripples are made in this pastime, are still able to run and hide, eluding the hunter even if he's willing to spend time searching for them, which he usually isn't.... *It's one thing to run down a cripple in a picked bean field or a pasture, and quite another to watch a wing-tipped bird drop into a huge block of switchgrass.* Oh nasty, nasty switchgrass. A downed bird becomes invisible on the ground and is practically unfindable without a good dog, and few "waterfowlers" have them these days. They're hard to train—usually a professional has to do it—and most hunters can't be bothered. Birds are easy to tumble.... *Canada geese—blues and snows—can all take a good amount of shot. Brant are easily called and decoyed and come down easily. Ruffed grouse are hard to hit but easy to kill. Sharptails are harder to kill but easier to hit....* It's just a nuisance to recover them. But it's fun, fun, fun swatting them down.... *There's distinct pleasure in watching a flock work to a good friend's gun....*

Teal, the smallest of common ducks, are really easy to kill. Hunters in the South used to *practice* on teal in September, prior to the "serious" waterfowl season. But the birds were so diminutive and the limit so low (four a day) that many hunters felt it hardly worth going out and getting bit by mosquitoes to kill them. Enough did, however, brave the bugs and manage to "harvest" 165,000 of the little migrating birds in Louisiana alone. *Shooting is usually best on opening day. By the second day you can sometimes detect a decline in local teal numbers. Areas may deteriorate to virtually no action by the third day. . . .* The area *deteriorates.* When a flock is wiped out, the skies are empty. *No action.* Teal have declined more sharply than any duck species except mallard; this baffles hunters. Hunters and their procurers—wildlife agencies—will never admit that hunting is responsible for the decimation of a species; instead they will deliver these familiar and litanic lines: *Hunting is not the problem. Pollution is the problem. Pesticides, urbanization, deforestation, and wetlands destruction is the problem.* And drought! There's been a big drought! Antis should devote their energies to solving these problems if they care about wildlife, and leave the hunters alone. While the Fish and Wildlife Service is busily conducting experiments in cause and effect, like releasing mallard ducklings on a wetland sprayed with the insecticide ethyl parathion (they died—it was known they would, but you can never have enough studies that show guns aren't a duck's only problem), hunters are killing some two hundred million birds and animals each year. But these deaths are considered incidental to the problem. A factor, perhaps, but a *minor* one. *Ducks Unlimited* says the problem isn't hunting, it's low *recruitment* on the part of the birds. To the hunter, birth in the animal kingdom is recruitment. They wouldn't want to use an emotional, sentimental word like *birth.* The black duck, a "popular" duck in the Northeast, so popular that game agencies felt that hunters couldn't be asked to refrain from shooting it, is scarce and getting scarcer. Nevertheless, it's still being hunted. *A number of studies are currently under way in an attempt to discover why black ducks are disappearing, Sports Afield* reports. Black ducks are disappearing because they've been shot out, their elimination being a dreadful example of game management and managers who are loath to "displease" hunters. The skies—*flyways*—of America have been divided into four administrative regions, and the states, advised by a federal government coordinator, have to agree on policies. A lot of squabbling always goes on in flyway meetings—lots of complaints about short-stopping, for example. Short-stopping is the deliberate holding of birds in a state, often by feeding them in wildlife refuges, so that their southern migration is slowed or stopped. Hunters in the North get to kill more than hunters in the South. This isn't fair. Hunters demand equity in opportunities to kill.

Wildlife managers hate closing the season on anything. Closing the season on a species would indicate a certain amount of mismanagement and misjudgment at the very least—a certain reliance on overly optimistic winter counts, a certain overappeasement of hunters who would be "upset" if they couldn't kill their favorite thing. And worse, closing a season would be considered victory for the antis. Bird hunting "rules" are very complicated, but they all encourage killing. There are shortened seasons and split seasons and special seasons for "underutilized" birds. (Teal were recently considered underutilized.) The limit on coots is fifteen a day—shooting them, it's easy! They don't fly high—knocking them down gives the hunter something to do while he waits in the blind for something better to come along. Some species are "protected," but bear in mind that hunters begin blasting away one half hour before sunrise and that most hunters can't identify a bird in the air even in broad daylight. Some of them can't identify birds in hand either, and even if they can (*#%'! I got me a canvasback, that duck's frigging protected . . .*), they are likely to bury unpopular or "trash" ducks so that they can continue to hunt the ones they "love."

Game "professionals," in thrall to hunters' "needs," will not stop managing bird populations until they've doled out the final duck (*I didn't get my limit, but I bagged the last one, by golly . . .*). The Fish and Wildlife Service services legal hunters as busily as any madam, but it is powerless in tempering the lusts of the illegal ones. Illegal kill is a monumental problem in the not-so-wonderful world of waterfowl. Excesses have always pervaded the "sport," and bird shooters have historically been the slobs and profligates of hunting. *Doing away with hunting would do away with a vital cultural and historical aspect of American life,* apologists claim. So, do away with it. Do away with those who have already done away with so much. Do away with them before the birds they have pursued so relentlessly and for so long drop into extinction, sink, in the poet Wallace Stevens's words, "downward to darkness on extended wings."

• • •

"Quality" hunting is as rare as the Florida panther. What you've got is a bunch of guys driving over the plains, up the mountains, and through the woods with their stupid license that cost them a couple of bucks and immense coolers full of beer and body parts. There's a price tag on the right to destroy living creatures for play, but it's not much. *A big-game hunting license is the greatest deal going since the Homestead Act,* Ted Kerasote writes in *Sports Afield. In many states residents can hunt big game* *for more than a month for about $20.* It's cheaper than taking the little woman out to lunch. It's cheap all right, and it's because killing animals is considered recreation and is underwritten by state and federal funds. In Florida, state moneys are routinely spent on "youth hunts," in which kids are guided to shoot deer from stands in wildlife management areas. The organizers of such events say that these staged hunts *help youth to understand man's role in the ecosystem.* (*Drop a doe and take your place in the ecological community, son. . . .*)

Hunters claim (they don't actually believe it, but they've learned to say it) that they're doing nonhunters a favor—for if they didn't use wild animals, wild animals would be useless. They believe that they're just *helping Mother Nature control populations* (you wouldn't want those deer to die of starvation, would you . . .). They claim that their tiny fees provide *all* Americans with wild lands and animals. (People who don't hunt get to enjoy animals all year round, while hunters get to enjoy them only during hunting season. . . .) *Ducks Unlimited* feels that it, in particular, is a selfless provider and environmental champion. Although members spend most of their money lobbying for hunters and raising ducks in pens to release later over shooting fields, they do save some wetlands, mostly by persuading farmers not to fill them in. *See that little pothole there the ducks like? Well, I'm gonna plant more soybeans there if you don't pay me not to. . . .* Hunters claim many nonsensical things, but the most nonsensical of all is that they pay their own way. They do not pay their own way. They do pay into a perverse wild-life management system that manipulates "stocks" and "herds" and "flocks" for hunters' killing pleasure, but these fees in no way cover the cost of highly questionable ecological practices. For some spare change . . . *the greatest deal going* . . . hunters can hunt on public lands—national parks, state forests—preserves for hunters!—which the nonhunting and antihunting public pay for. (Access to private lands is becoming increasingly difficult, as experience has taught people that hunters are obnoxious.) Hunters kill on millions of acres of land all over America that is maintained with general taxpayer revenue, but the most shocking, really twisted subsidization takes place on national wildlife refuges. Nowhere is the arrogance and the insidiousness of this small, aggressive minority more clearly demonstrated. Nowhere is the murder of animals, the manipulation of language, and the distortion of public intent more flagrant. The public perceives national wildlife refuges as safe havens, as sanctuaries for animals. And why wouldn't they? The word refuge means shelter from danger and distress. But the foolish nonhunting public—they tend to be so literal. The word has been reinterpreted by management over time, and now hunters are invited into more than half of the country's more than 440 wildlife "sanctuaries" each year to bang them up and kill more than half a million animals. This is called *wildlife-oriented recreation.* Hunters think of this as being no less than their due, claiming that refuge lands were purchased with duck stamps (. . . *our duck stamps paid for it . . . our duck stamps paid for it*). Hunters equate those stupid stamps with the mystic, multiplying power of the Lord's loaves and fishes, but of ninety million acres in the Wildlife Refuge System, only three million were bought with hunting-stamp revenue. Most wildlife "restoration" programs in the states are translated into clearing land to increase deer habitats (*so that too many deer will require hunting . . . you wouldn't want them to die of starvation, would you?*) and trapping animals for restocking and redistribution (so more hunters can shoot closer to home). Fish and game agencies hustle hunting. It's time for them to get into the business of protecting and preserving wildlife and creating balanced ecological systems instead of pimping for hunters who want their deer/duck/pheasant/turkey—animals stocked to be shot.

Hunters' self-serving arguments and lies are becoming more preposterous as nonhunters awake from their long, albeit troubled sleep. Sport hunting is immoral; it should be made illegal. Hunters are persecutors of nature who should be prosecuted. They wield a disruptive power out of all proportion to their numbers, and pandering to their interests—the special interests of a group that just wants to kill things—is mad. It's grotesque that every year less than 7 percent of the population turns the skies into shooting galleries and the woods and fields into abattoirs. It's time to stop actively supporting and passively allowing hunting, and time to stigmatize it. It's time to stop being conned and cowed by hunters, time to stop pampering and coddling them, time to get them off the government's duck-and-deer dole, time to stop allowing hunting to be creditable by calling it "sport" and "recreation." Hunters make wildlife *dead, dead, dead.* It's time to wake up to this indisputable fact. As for the hunters, it's long past checkout time.

Cabin Cabin

TED'S CABIN WAS 10 X 12. HE BUILT IT HIMSELF. WHEN HE wanted to go to town, which was Lincoln, Montana, which was five miles away, he pedaled there on an old bicycle. For errands farther away, as we know, he relied on public transportation. Ted's cabin is intriguing. Plucked from its beautiful forested site near Montana's Scapegoat Wilderness, which is near the Bob Marshall Wilderness, which is near nothing, it was taken to reside in a stark hangar at a former Air Force base in Sacramento, California. What do you think it's thinking there, stored. It's thinking, *Where is he, where's all my stuff, the 232 books, those typewriters, that manuscript, those little devices he was always fiddling with?* It's thinking, *I harbored a fugitive. I'm an enabler, I'm an accomplice. I'm going to be sentenced and reduced to kindling, packed in little boxes and sacks and sold, the proceeds going to the victims' families.* The cabin's thinking, *I'm going to burn!*

The cabin, all wrapped like a Christo and transported to Sacramento on a flatbed truck, was to be used by the Unabomber defense in the case before the jurors. The argument would be that no one but a lunatic could live in such a primitive dwelling for so many years. Eighteen years! A 10 x 12 shack! Two little windows. No plumbing or electricity.

63

Cozy, the cabin thinks. *And it was lovely outside. Anyway, I never asked to be built.*

But the jurors never saw the cabin; they scarcely saw Ted. For Ted never got his day in court. He pleaded guilty in exchange for a life sentence because he did not want to go through a trial where his defense team would argue that he was insane, a "sickie" in Ted's word.

The victims' families said they wanted Ted to rot in hell, the cabin thinks. *Mean.*

The only place in Lincoln that is still selling Unabomber T-shirts is a boutique attached to the town's Exxon station. Now, some people, oh just call us loony, never patronize Exxon stations because of Alaska's Prince William Sound. One of the Unabomber's targets was the ad company that managed to sanitize the corporation's image after the disaster of a decade ago when the tanker, the *Exxon Valdez,* spilled eleven million gallons of crude that fouled thirteen hundred miles of shoreline and killed a quarter of a million birds and twenty-eight hundred sea otters.

They still haven't paid the five billion dollars they were supposed to, the cabin muses. *They said paying it would send a "perverse message."*

So this is a dilemma if you really want a Unabomber T-shirt but you've got principles too. The T-shirt has a black drawing of the cabin on it and says "Mountain Cabin Hideout of the Unabomber."

Not a hideout, a home, the cabin thinks.

There's a splotch of red, denoting Ted's bicycle leaning against the cabin, which is a nice touch. *Where's that silly bicycle now?* the cabin thinks. The library, which Ted frequented to read *Omni* and *Scientific American,* never sold T-shirts. If they had, they'd have a new research wing by now. *He read Thackeray and Shakespeare too.* The cabin remembers *"The friend hath lost his friend."* That's from King Lear.

The cabin thinks, *I agree with the Unabrother. Ted is a complex man.*

• • •

Ted never claimed to be a great writer, but at least he didn't giddily anthropomorphize WILD Nature, which he loved and considered the corrective to technology's horrid habits. "Nature, that which is outside the power of the system, is the opposite of technology which seeks to expand indefinitely the power of the system" (The *Manifesto*, paragraph 184). The *Manifesto* was 35,000 words long when it ran in the *Washington Post* in 1995 and was described by an editor there as a "romantic, turgid, disturbing document." But it wasn't all that bad to read. Ted thought he had to be reasonable in his writing, unemotional and precise. That other famous American cabin builder, Henry David Thoreau, thought he could get away with anything in his prose. "I also heard the whooping of the ice in the pond, my great bedfellow in that part of Concord, as if it were . . . troubled with flatulence and bad dreams. . . ." He wrote that in *Walden.* Henry could be silly. Too, Nature was a business for Henry, an occupation, and his cabin-in-the-woods experiment has become one of the most overinflated of American myths. Walden Pond in Concord, Massachusetts, was a simulated wilderness even back in 1845. The cabin was in view of the public road and its scribbling occupant had a constant stream of visitors. It wasn't as though he had nothing but a farting pond for company. And he lived there for only two years before returning to the gabby salons of town. Henry estimated the cost of his cabin at $28.12 and went into great detail about its construction in his tiresome way so that there are hundreds, perhaps even thousands of clone cabins around today exactly like his, exactly 15 x 10 with brick fireplace, shingled roof and sides, wide board doors, and root cellar.

No no no no, the cabin thinks. *No clone we. We were unique.*

Massachusetts cabin. The term has no cachet. It conjures a blank. Massachusetts cabin. Nothing. Now, a Maine cabin. People in and around a Maine cabin have cocktails by the flagpole, varnish the twig furniture, and bake blueberry muffins by the canoe load. Occupants of Alaska cabins carve fetishes out of moose droppings, listen to the permafrost melt, and dream about what they're going to do with the money the state pays them each year just to live in Alaska. Idaho cabin . . . Now, this may not be fair, but I think of people just succumbing in Idaho cabins. I think of unpleasant accommodations being made. Remember the Hemingway story—it was one of the Nick Adams stories—where the peasant's wife dies in the middle of winter and he props her up in the woodshed because the ground's too frozen to bury her and he uses her open mouth to hang his lantern from every time he goes out to gather wood? That didn't happen in Idaho, admittedly. It happened in the Italian Alps. And not only did it not happen in Idaho, it happened fictionally. But the story still possesses what Idaho cabin suggests. A glumness. (There is, of course, Papa's final irrefutable association with Idaho.) I see all those Idaho cabins packed with potatoes and ready to serve shotguns and ennui. An Idaho cabin is just slightly off cabinwise, as is an Ozark cabin. But a Montana cabin ... that is something other, way other. A Montana cabin gives shelter and substance to wild thought.

I always thought I was his muse, but he's still writing. Writing without me. Five hundred and forty-eight pages to be exact. In a prison cell yet.

In May of 1998 Theodore Kaczynski was sentenced to four life sentences. (Don't you ever wonder how that kind of immortality conferred *works?*) He had wanted to present a defense based on his views about technology and the environment, but his efforts to represent himself or be represented by a lawyer who would argue his chosen defense were rejected by the judge. In the fall of 1999, however, a federal appeals court agreed to hear his request to withdraw his guilty plea and receive a juried trial, even though this might very well result in a death sentence.

He always wanted to be heard. This can't be existence, being stored. We're both being stored!

His previous lawyers had arranged for a psychiatric examination without his consent, an examination that concluded that he was fit to stand trial even though he was a "sickie," or in more psychologically precise terms, a paranoid schizophrenic awash in delusions—the worst one being that technology is the vehicle by which people are destroying themselves and the world.

What? It's not true?

Neverglades

THAT THE EVERGLADES STILL EXISTS IS A COLLECTIVE ILLU-sion shared by both those who care and those who don't. People used to say that nothing like the Everglades existed anywhere else in the world, but it doesn't exist in South Florida anymore either. The Park, which millions of people visit and perceive to be the Everglades, makes up only 20 percent of the historic Glades and is but a pretty, fading afterimage of a once astounding ecosystem, the remaining 80 percent of which—drained, diked, and poisoned—has vanished beneath cities, canals, vast water impoundment areas, sugarcane fields, and tomato farms. Ninety percent of the wading bird population has disappeared in fifty years, and gradually (quickly) "one of the rarest places on earth" (as it is so frequently described) located conveniently (unfortunately) one hour from Miami, has become a horror show of extirpated species. On land, a water park with no water; at sea, a sick marine estuary turning into a murky, hyper-saline, superheated lagoon.

The Everglades watershed once began just south of Orlando with the Kissimmee River winding 103 miles to Lake Okeechobee. The lake, a vast but shallow depression then spilled the waters through seepage, springs, and overflow across and down the entire peninsula, eventually passing into

Florida Bay and the Gulf of Mexico. After a hurricane in 1928 flooded land drained for agriculture and drowned more than a thousand migrant workers, the Army Corps of Engineers was directed to build a dike, a thirty-foot-high barrier that encircled the lake. A vast 700,000-acre area to the south—more than one-third of the Everglades watershed but no longer and never to be again the Everglades—became the Everglades Agricultural Area. East and south of the EAA (all planted in sugarcane, a simple crop with a simple preference—dry land). Vast portions of the Everglades (never to be again the Everglades) were turned into Water Conservation Areas, lifeless holding tanks compartmentalized by canals. "Excess" water was flushed out to sea.

In 1961, the Army Corps of Engineers was directed to channelize the Kissimmee. (Kissimmee is said to mean "heaven's gate" in the language of the Calusa Indian, Florida's original and long-extinct Indian tribe. Since extremely little is known about the Calusa, this pretty notion is unlikely, though an inarguable fact is that the town of Kissimmee, a bawdy sprawl of billboards, flashing signs, motels, and dinner theaters, no longer marks the ecological beginnings of the mysterious Everglades, but is the gateway to the fantasy lands of Disney World.) It took ten years to transform the river (which wandered a mile east or west for every mile it flowed south) into a fifty-two-mile straight-edged canal two hundred feet wide and thirty feet deep. Forty-five thousand acres of wetlands dried up, the wildlife vanished, and a slug of pollutants, mostly cow shit from the dairy farms that moved in after the ditch was dug, directly entered Lake Okeechobee, which was already gagging on nutrient overload from Big Sugar's back-pumped used irrigation water. (It's hard to believe the power and influence Sugar has, but think of the NRA with a sweet tooth.) Lake O. sat fat in the center of peninsula Florida like a nasty blood-swollen tick.

By the '70s, the elaborate plumbing system had been completed and was rapidly accomplishing its engineered purpose of sucking the Everglades dry. (The Corps even deliberately introduced an alien tree—the melaleuca. From Australia, a sort of vampirish member of the myrtle family, the melaleuca sucks up enormous amounts of water, grows fast and thickly, and because it flowers five times a year, spreads rampantly.) At this point, uncertainty arose as to the excellence of this idea. The Park (at the end of the pipe) was supposed to exist as a museum piece, not connected with the commerce of the real world, but the Everglades, which was the actual world of South Florida, had become so depleted of its original abundance and ecological function that it was no longer the Everglades at all. The gentle, natural, rain-driven sheet flow that once sustained it had been replaced by erratic pulses of water, which came in gorged polluted flushes, too much or too little, and always in the wrong season.

The Everglades was lost, but it was still there and the notion took hold that with goodwill and a little tinkering it could be resurrected. To some degree. The public became increasingly educated about its intrinsic worth. For example, around this time a group of sixth-graders visiting the Park composed a poem that was rendered on a bronze plaque and erected at the Pay-Hay-Okee boardwalk overlook.

Every time you go to a place
That has those animals on its face,
It makes you laugh and cheer
Because it's fun out here.
We love you Everglades!
We'll help to save the place!
Nice work, children! Lovely.

The Everglades, no longer quite existing but still troublingly existent, was increasingly being deemed worthy of love, of being saved. Studies commenced. Debate ensued over water timing and quantity. (Concerns over water quality were tabled for later.) Officials were comfortable with the debate-and-study process, which went on pretty much for a generation. The Everglades could be fixed if only further studies could be conducted and as long as any conclusions and recommendations accommodated agricultural needs and rising population demands. In concert with the Everglades' increasing degradation and diminishment, more and more acts were passed to protect it. At the same time Bob Graham, a two-term governor of the state and now a US senator, was establishing a task force called Save Our Everglades, he was also pushing for increased sugar subsidies. While water management was dribbling water into the starved Shark River Slough, heart of the Park, in an "experiment," it was accelerating the dumping of hundreds of millions of gallons into the Atlantic because that was the way the plumbing system was set up. At the same time Congress enacted the Everglades National Park Protection and Expansion Act, authorizing the addition of 107,000 acres on the eastern side of the Park, the Corps of Engineers was proposing more canals to protect tomato interests nearby and provide drainage for two hundred homes in the area that the government was in the process of buying. At the same time Lake O. was being cleaned up because Sugar was no longer pumping dirty irrigation water into it, the Park was suffering because Sugar had redirected effluent there. At the same time the federal government was insisting that "natural hydrological conditions be restored" to the Everglades, Sugar was flushing two hundred tons of phosphorous into the wetlands each year, replacing the sawgrass "face" of the Glades with cattails and stinking organic ooze. At the same time the Kissimmee River was being considered eligible for restoration (the first step was restoring its name, which had been changed to the less evocative C-38), Florida Bay's waters, once crystalline, were turning murky and opaque, a milky green, a pea green, even an uncanny phosphorescent green. Cape Sable, once known for its wild white empty shore, was becoming remarkable because it signifies the beginning of the Dead Zone, a spreading area of massive turtle grass die-off that has fueled an algae bloom in which marine life perishes or from which it flees. It was a death soup, like. Florida Bay—terminus of the Everglades.

And still, throughout the 1980s, the appellations kept coming. The Everglades became a World Heritage Site, an International Biosphere Preserve, a Wetland of International Significance. The sicker it became, the more the perception was that its condition was being recognized and help was on the way. By the 1990s, the Everglades was—or was going to be—the test case for proving that the ecosystems of America could be protected. Bruce Babbitt, the secretary of the interior, flew over Florida Bay and was "appalled." He vowed to make the Everglades his Number 1 priority, it was to be the keystone of the Clinton administration's environmental policy. Cooperation was necessary, of course, a comprehensive plan would be required, and coexistence among parties of radically different objectives would have to be encouraged. (Build those wildlife tunnels under the new four-lane highways and make those panthers use them!) After decades of chatter about the retooling, restoration, replication, reconstruction, and repair of the Everglades, something was at last about to happen. A *Statement of Principles* was issued. Success would be achieved by "an unprecedented new partnership, joining the Federal and State governments with the agricultural interests of South Florida"—in other words, the exact same alliance that had been suffocating the Everglades all along. The Everglades Forever Act was signed, capping the sugar industry's clean-up costs, delaying water-quality standards, and limiting future efforts at restoring water flow to areas not employed by sugar. Still, defenders of the Everglades remained hopeful, for a final study was still to be done, the Corps' restudy, which was expected to recommend drastic changes in water conservation and delivery. Yet even the most myopic and optimistic friend to the Everglades should have felt a twinge of concern when Alfonso Fanjul, Sugar's billionaire, appeared in Special Prosecutor Kenneth Starr's report on President Clinton. The president interrupted his dalliance with Monica only once, to take a call from Fanjul. This was troubling, environmentally speaking, this implication of respectful, very respectful acquaintance with Fanjul but ... but ... eco-carers mustn't be cynical—otherwise, they'd give up, they'd just stay in their nest and drown—so the phone call was put in . . . perspective.

In the summer of '99, just in time for the future, the Clinton administration presented the final Everglades restoration plan to Congress. It looked suspiciously like a massive water supply project for Florida development. The Everglades would still be at the mercy of cane and commerce; only 240 miles of the more than 1,800 miles of levees and canals will be removed, and though less water will be thrown to the sea, the murderous manmade cycles of drought and flood will remain in place. The final plan—costing $8 billion—the final solution locks in the ultimate nonsolution. And they couldn't have done it without us, the children of all ages, naive and hopeful to the bone. Couldn't have done it without our patience through the many many years, years that Sugar utilized by perfecting their lobbying powers and building an immense refinery on site, and that developers exploited with a paroxysm of blanket building on both coasts. All those years. . . . Looking back on it, we might just as well have brought a gallon of water with us when we visited the Park and dumped it on the ground.

We love you Everglades!
We'll help to save the place!

Chumps.

Florida

FLORIDA HAS LONG BEEN CONSIDERED A PLAYGROUND, A
location set for vast theme parks, a sunny eccentricity, a beach.
She was by no means America's last frontier, but still in the
1880s, the major part of the state was unexplored wilderness,
wetlands and swamps, immense piney flatwoods, live oak and
cypress forests, twisting rivers and impenetrable mangrove
mazes, with everywhere strange creatures, giant reptiles and
great flocks of resplendent and remarkable birds. Key West,
the tiny coral island more than one hundred miles off the tip
of peninsular Florida was then improbably the largest, wealth-
iest, and most sophisticated city of this vast and improbable
state. Miami and Palm Beach did not exist. Florida's natural
wonders—her fishes and reefs and palmy beaches—seemed
extraordinarily exotic. She was simply unlike anywhere else,
and after she was tamed and drained, entrepreneurs could
make her appear to be any number of things. Florida could
be anywhere and serve the fantasies of anyone. She could
become the repository for any number of crafted and imagi-
nary histories, a wealth of simulated habitats and instant
tropical gardens. She could be the African veld, the Mediter-
ranean, Venice, and despite hurricanes, frosts, and mosquitoes
so plentiful it was said that a person could come up with a

quart of them in a pint jar, she was fashioned into all these contrivances and more.

The ordinary citizen flocked to Florida because he was sure that he would be entertained, amused. The invention of Florida resulted in the invention of modern tourism. Florida became both outlandish and quaint with her alligator wrestling and lion farms, her mermaid shows and parrot jungles. Everything from coconuts to cypress knees to coral became a souvenir. The very sunshine was employed as a huckster, as well as that most fragrant of scents, the orange blossom. The flamingo, an extravagantly vivid bird that scarcely looks real (which is not to say that it doesn't look exactly like a flamingo) became the very image of a colorfully extravagant state (even though it is a native of the Caribbean and not a Florida bird at all). Everything in Florida was perceived to be, and marketed as being, curious. Grapefruits were very curious. As were the banyan trees. There were fossilized sharks' teeth on the beach. There were pelicans. Everything seemed prehistoric and slightly preposterous.

Florida is a show, an entire state dedicated to the vacation principle. And as Florida becomes more and more what she has become, a state attuned to growth, on autocatalytic open throttle, the pace quickens to promote "hidden" Florida. On the west coast near Cedar Key, counties less developed are referred to in tourist development literature as "Mother Nature's Theme Park." "Real" Florida becomes a packaged experience. In the Keys, visitors pay to swim with dolphins, the theory being that you will appreciate the mammal more because of it. The dolphin has evolved from being just another animal entertainer to being the supreme educational showman of the natural world. (The manatee, a dear and improbable creature and, next to the panther, the most endangered in the state, has too modest and retiring a personality to be employed professionally in an "interactive" capacity. The rank roadside zoos of Florida—*Mother Nature's Creatures on Parade*, once a tourism staple—have mostly gone or been transmogrified into a format presented on a somewhat different plane—the wildlife rehabilitation center. Here the tourist (why there were 40 million of you just last year) or the resident (why there are 15 million of us in place just now) can view unlucky but still engaging wildlife in an airy natural setting, and know they are being kept viable by the most determined volunteer help available. Increasingly popular with the kids and the caring, wildlife rehabilitation centers have even become a plus tourist-wise, their broken guests educational totems that remind us of the impact our sheer numbers have on their world, of the difficulties, even improbability, of sharing Florida with them. Even the broken and maimed can be viewed positively in sunny, entertaining Florida.

The Case against Babies

Babies, babies, babies. There's a plague of babies. Too many rabbits or elephants or mustangs or swans brings out the myxomatosis, the culling guns, the sterility drugs, the scientific brigade of egg smashers. Other species can "strain their environments" or "overrun their range" or clash with their human "neighbors," but human babies are always welcome at life's banquet. Welcome, Welcome, Welcome—Live Long and Consume! You can't draw the line when it comes to babies because ... where are you going to draw the line? *Consider having none or one and be sure to stop after two*, the organization Zero Population Growth suggests politely. Can barely hear them what with all the babies squalling. Hundreds of them popping out every minute. Earth's human population has more than tripled in the last century. Ninety-seven million of them each year. While legions of other biological life forms go extinct (or, in the creepy phrase of ecologists, "wink out"), human life bustles self-importantly on. Those babies just keep coming! They've gone way beyond being "God's gift"; they've become entitlements. Everyone's having babies, even women who can't have babies, particularly women who can't have babies—they're the ones who sweep fashionably along the corridors of consumerism with their double-wide strollers, stuffed with twins and triplets. (Women push those things with the effrontery of someone piloting a bulldozer, which strollers uncannily bring to mind.) When you see twins or triplets, do you think, *aw* or *ooh* or *that's sort of cool, that's unusual,* or do you think, *That woman dropped a wad on in-vitro fertilization, twenty-five, thirty thousand dollars, at least. . . .*

The human race hardly needs to be more fertile, but fertility clinics are booming, making millionaires of the hot-shot fertility doctors who serve anxious gottahavababy women, techno-shamans who have become the most important aspect of the baby process, giving women what they want: BABIES. (It used to be a mystery what women wanted, but no more . . . Nietzsche was right. . . .) Ironically—though it is far from being the only irony in this baby craze—women think of themselves as being *successful, personally fulfilled* when they have a baby, even if it takes a battery of men in white smocks and lots of hormones and drugs and needles and dishes and mixing and inserting and implanting to make it so. Having a baby means *individual completion* for a woman. What do boys have to do to be men? Sleep with a woman. Kill something. Yes, killing anything large-ish in the animal kingdom, or even another man, appropriate in times of war, has ushered many a lad into manhood. But what's a woman to do? She gets to want to have a baby.

While much effort has been expended in third world countries educating women into a range of options that does not limit their role merely to bearing children, well-off, educated, and indulgent American women are clamoring for babies, babies, BABIES to complete their status. They've had it all, and now they want a baby. And women over thirty-five want them NOW. They're the ones who opt for the aggressive fertility route; they're impatient; they're sick of being laissez-faire about this. Sex seems such a laborious way to go about it. At this point they don't want to endure all that intercourse over and over and maybe get no baby. What a waste of time! And time's awasting. *A life with no child would be a life perfecting hedonism,* a forty-something infertile woman said, now the proud owner of pricey twins. Even women who have the grace to submit to fate can sound wistful. *It's not so much that I wish that I had children now,* a travel writer said, *but that I wish I had had them. I hate to fail at anything.* Women are supposed to wish and want and not fail.

• • •

The eighties were a decade when it was kind of unusual to have a baby. Oh, the lower classes still had them with more or less gusto, but professionals did not. Having a baby was indeed so quaintly rebellious and remarkable that a publishing niche was developed for men writing about babies, *their* baby, their baby's first year in which every single day was recorded (he slept through the night . . . he didn't sleep through the night . . .). The writers would marvel over the size of their infant's scrotum; give advice on how to tip the obstetrician (not a case of booze, a clock from Tiffany's is nicer); and confess that their baby exhibited intelligent behavior like rolling over, laughing, and showing fascination with the TV screen far earlier than normal children. Aside from the talk about the poopie and the rashes and the cat's psychological decline, these books frequently contained a passage, an overheard bit of Mommy-to-Baby monologue along these lines: *I love you so much I don't ever want you to have teeth or stand up or walk or go on dates or get married, I want you to stay right here with me and be my baby.* . . . Babies are one thing. Human beings are another. We have way too many human beings. Almost everyone knows this.

Adoption was an eighties thing. People flying to Chile, Russia, Guatemala, all over the globe, returning triumphantly with their BABY. It was difficult, adventurous, expensive, and generous. It was trendy then. People were into adopting bunches of babies of all different cultures, one of each. Adoption was a fad, just like the Cabbage Patch dolls, which fed the fad to tens of thousands of prepubescent girl consumers.

Now it is *absolutely* necessary to digress for a moment and provide an account of that marketing phenomenon. These fatuous-faced soft-sculpture dolls were immensely popular in the eighties. The gimmick was that the dolls were "born," you couldn't just buy the damn things. If you wanted one, you had to "adopt" it. Today they are still being born and adopted, although at a slower rate, in Babyland General Hospital, a former medical clinic right on the fast-food and car-dealership strip in the otherwise unexceptional north Georgia town of Cleveland. There are several rooms at Babyland General. One of them is devoted to the preemies (all snug in their little gowns, each in its own spiffy incubator), and another is devoted to the cabbage patch itself, a suggestive mound with a fake tree on it from which several times a day comes the announcement CABBAGE IN LABOR! A few demented moments later, a woman in full nurse regalia appears from a door in the tree holding a brand-new Cabbage Patch Kid by the feet and giving it a little whack on the bottom. All around her in the fertile patch are happy little soft heads among the cabbages. Each one of these things costs $175, and you have to sign papers promising to care for it and treasure it forever. There are some cheesy dolls in boxes that you wouldn't have to adopt, but children don't want those—they want to sign on the line, want the documentation, the papers. The dolls are all supposed to be different, but they certainly look identical. They've got tiny ears, big eyes, a pinched rictus of a mouth, and lumpy little arms and legs. The colors of the cloth vary for racial verisimilitude, but their expressions are the same. They're glad to be here, and they expect everything.

But these are just dolls, of course. The *real* adopted babies who rode the wave of fashion into many hiply caring homes are children now, an entirely different kettle of fish, and though they may be providing (just as they were supposed to) great joy, they are not darling babies anymore. A baby is not really a child; a baby is a BABY, a cuddleball, representative of virility, wombrismo, and humankind's unquenchable wish to outfox Death.

Adoptive parents must feel dreadfully dated these days. Adoption—how foolishly sweet, so kind of naive. With adopted babies, you just don't know, it's too much of a crap shoot. Oh, they *told* you that the father was an English major at Yale and that the mother was a brilliant mathematician and harpsichordist who was just not quite ready to juggle career and child, but what are you going to think when the baby turns into a kid who is trying to drown the dog and set national parks on fire? Adoptive parents do their best, of course, at least as far as their liberal genes allow; they look into the baby's *background*, they don't want just any old baby (even going to the dog and cat pound, you'd want to pick and choose, right?); they want a pleasant, healthy one, someone who will appreciate the benefits of a nice environment and respond to a nurturing and attentive home. They steer away (I mean, one has to be realistic, one can't save the world) from the crack and smack babies, the physically and mentally handicapped babies, the HIV and fetal-alcohol syndrome babies.

Genes matter, more and more, and adoption is just too . . . where's the connection? Not a single DNA strand to call your own. Adoption signifies that you didn't do everything you could; you were too cheap or shy or lacked the imagination to go the energetic fertility route, which, when successful, would come with the assurance that some part of the Baby or Babies would be a continuation of you, or at the very least your companion, loved one, partner, whatever.

• • •

I once prevented a waitress from taking away my martini glass, which had a tiny bit of martini remaining in it, and she snarled, "Oh, the precious liquid," before slamming it back down on the table. It's true that I probably imagined that there was more martini in the glass than there actually was (what on earth could have happened to it all?), but the precious liquid remark brings unpleasantly to mind the reverent regard in which so many people hold themselves. Those eggs, that sperm, oh precious, precious stuff! There was a terrible fright among humankind recently when some scientists suggested that an abundance of synthetic chemicals was causing lower sperm counts in human males—awful, awful, awful—but this proves not to be the case; sperm counts are holding steady and are even on the rise in New York. Los Angeles males don't fare as well (do they drink more water than beer?), nor do the Chinese, who, to add insult to insult, are further found to have smaller testicles, a finding that will undoubtedly result in even more wildlife mutilation in the quest for aphrodisiacs. Synthetic chemicals do "adversely affect" the reproductive capabilities of nonhuman creatures (fish, birds), but that is considered relatively unimportant. It's human sperm that's held in high regard, and in this overpopulated age it's become more valuable—good sperm, that is, from intelligent, athletic men who don't smoke, drink, do drugs, have AIDS or a history of homicide—because this overpopulated age is also the donor age. Donor sperm, donor womb, donor eggs. Think of all the eggs that are lost to menstruation every month. The mind boggles. Those precious, precious eggs, lost. (Many egg donors say they got into the business because they didn't like the idea of their eggs "going to waste.") They can be harvested instead and frozen for a rainy day or sold nice and fresh. One woman interviewed in the *New York Times* has made it something of a career. *I'm not going to just sit home and bake cookies for my kids. I can accomplish things,* she says. No dreary nine-to-five desk job for her. She was a surrogate mother for one couple, dishing up a single baby; then she donated some eggs to another couple, who had a baby; now she's pregnant with twins for yet another couple. *I feel like a good soldier, as if God said to me, 'Hey, girl, I've done a lot for you, and now I want you to do something for Me,'* this entrepreneurial breeder says. (It's sort of cute to hear God invoked, sort of for luck, or out of a lingering folksy superstition.) Egg donors are regular Jenny Appleseeds, spreading joy, doing the Lord's work, and earning a few bucks all at once, as well as attaining an odd sense of empowerment (I've got a bunch of kids out there, damned if I know who they all are ...).

One of the most successful calendars published each year is Anne Geddes's BABIES. Each month shows the darling little things on leaves in the rhubarb patch, cupped in a tulip, as little bees in a honeycomb, and so on—solemn, bright-eyed babies. They look a little bewildered though, and why shouldn't they? How did they get here? They were probably mixed up in a dish. Donor eggs (vacuumed up carefully through long needles); Daddy's sperm (maybe ... or maybe just some high-powered NYC dude's); gestational carrier; the "real" mommy waiting anxiously, restlessly on the sidelines (want to get those babies home, start buying them stuff). Baby's lineage can be a little complicated in this one big worldwebby family. With the help of drugs like Clomid and Metrodin and Perganol there are an awful lot of eggs out there these days—all being harvested by those rich and clever, clever doctors in a "simple procedure" and nailed with bull's-eye accuracy by a spermatozoon. One then gets to "choose" among the resulting cell clumps (or the doctor gets to choose; he's the one who knows about these things), and a number of them (for optimum success) are inserted into the womb, sometimes the mother's womb and sometimes not. These fertilized eggs, unsurprisingly, often result in multiple possibilities, which can be decreased by "selective reduction." They're not calendar babies yet, they're embryos, and it is at this point, the multiple-possibility point, that the mother-to-be often gets a little overly ecstatic, even greedy, thinking ahead perhaps to the day when they're not babies any longer, the day when they'll be able to amuse themselves by themselves like a litter of kittens or something—if there's a bunch of them all at once, there'll be no need to go through that harrowing process of finding appropriate playmates for them. She starts to think, *Nannies probably don't charge that much more for three than for two*, or *Heaven knows we've got enough money or we wouldn't have gotten into all this in the first place*. And many women at the multiple-possibility point, after having gone through pretty much all the meddling and hubris that biomedical technology has come up with, say demurely, *I don't want to play God* (*I DON'T WANT TO PLAY GOD?*) or *It would be grotesque to snuff one out to improve the odds for the others* or *Whatever will be will be*.

So triplets happen, quads and quints and sextuplets, even *septuplets*. That Iowa fellow, that billing clerk at the Chevy dealership, Kevin McCaughey? said, *God could have given us one, but God decided to give us seven*. Well . . . not exactly. Mrs. McC (or rather Mrs. McC's ovaries) clearly overreacted to the drug she was administered, at which point Mr. McC's sperm could have been sensibly withheld and added into the mix in another month when, dosage adjusted, fewer eggs would have been produced. But, too late for that now, they're here— Brandon and Alexis and Natalie and Kelsey and Kenneth and . . . and . . . I forget—going through 350 diapers a week. And as soon as they—and all the other multiples too, even the less prestigious single baby—are old enough to toddle into day care, you can be assured that they'll be responsibly taught the importance of their one and only Earth, taught the 3Rs: Reduce, Reuse, Recycle. Too many people (which is frequently considered undesirable—gimme my space!) is caused by too many people (it's only logical), but it's mean to blame the babies, you can't blame the babies, they're innocent. Those poor bean counters at the United Nations Population Fund say that at current growth rates, the world will double its population in forty years. Overpopulation poses the greatest threat to all life on earth, but most organizations concerned with this problem don't like to limit their suggestions to the most obvious one—DON'T HAVE A BABY!—because it sounds so negative. Instead, they provide additional, more positive tips for easing the pressures on our reeling environment, such as car pooling and tree planting. (A portion of the proceeds from that adorable best-selling BABIES calendar goes to the Arbor Day Foundation for the planting of trees.)

• • •

Some would have it that not having a baby is *disallowing* a human life, horribly inappropriate in this world of rights. Everyone has rights; the unborn have rights; it follows that the *unconceived* have rights. (Think of all those babies pissed off at the fact that they haven't even been thought of yet.) Women have the right to have babies (we've fought so hard for this), and women who can't have babies have an even bigger right to have them. These rights should be independent of marital or economic status, or age. (Fifty- and sixty-something moms tend to name their babies after the gynecologist.) The reproduction industry wants fertility treatments to be available to

ILL NATURE

anyone and says that it wouldn't all be so expensive if those recalcitrant insurance companies and government agencies like Medicare and Medicaid weren't so cost-conscious and discriminatory and would just cough up the money. It's not as though you have to take out a *permit* to have a baby, be *licensed* or anything. What about the rights of a poor, elderly, feminist cancer patient who is handicapped in some way (her car has one of those stickers . . .) who wants to assert her right to independent motherhood and feels entitled to both artificial insemination into a gestational "hostess" and the right to sex selection as a basis for abortion should the fetus turn out to be male when she wants a female? Huh? What about her? Or what about the fifteen-year-old of the near future who kind of wants to have her baby even though it means she'll be stuck with a kid all through high school and won't be able to go out with her friends anymore who discovers through the wonders of amniocentesis and DNA analysis that the baby is going to turn out fat, and the fifteen-year-old just can't deal with fat and shouldn't have to . . . ? Out goes the baby with the bathwater.

But these scenarios are involved merely with messy political or ethical issues, the problematical, somewhat gross by-products of technological and marketing advances. Let the philosophers and professional ethicists puzzle over this and let the baby business boom. Let the courts figure it out. Each day brings another more pressing problem. Implanted with their weak-cervixed daughter's eggs and their son-in-law's sperm, women became pregnant with their own grandchildren; frozen embryos are inadvertently thawed; eggs are pirated; eggs are harvested from aborted fetuses; divorced couples battle

88

over the fate of cryopreserved material. "We have to have better regulation of the genetic product—eggs, sperm, and embryos—so we can legally determine who owns what," a professor of law and medicine at a California university says plaintively. (Physicians tend to oppose more regulation, however, claiming that it would impede research.)

• • •

While high-tech nations are refining their options eugenically and quibbling litigiously, the inhabitants of low-tech countries are just having babies. The fastest growth in human numbers in all history is going to take place in a single generation, an increase of almost five billion people (all of whom started out as babies). Ninety-seven percent of the surge is going to take place in developing countries, with Africa alone accounting for 35 percent of it (the poorer the country, the higher the birth rate, that's just the way it is). These babies are begotten in more "traditional," doubtless less desperate ways, and although they are not considered as fashion statements, they're probably loved just as much as upper-class western babies (or that singular one-per-family Chinese boy baby) and are even considered productive assets when they get a little older and can labor for the common good of their large families by exploiting more and more, scarcer and scarcer resources.

The argument that western countries with their wealth and relatively low birth rate do not fuel the population crisis is, of course, fallacious. France, as national policy, urges its citizens to procreate, giving lots of subsidies and perks to those French who make more French. The US population is growing faster than that of eighteen other industrialized nations, and, in terms of energy consumption, when an American couple stops spawning at two babies, it's the same as an average East Indian couple stopping at sixty-six, or an Ethiopian couple drawing the line at one thousand.

Yet we burble along, procreating. We're in a baby glut, yet it's as if we've just discovered babies, or invented them. Reproduction is sexy. Assisted reproduction is cool. The announcement that a movie star is going to have a baby is met with breathless wonder. A BABY! Old men on their third marriage regard their new babies with "awe" and crow about the "ultimate experience" of parenting.

It's as though, all together, in the waning years of the dying century, we collectively opened the Door of our Home and instead of seeing a friend standing there in some sweet spring twilight, someone we had invited over for drinks and dinner and a lovely civilized chat, there was Death, with those creepy little black seeds of his for planting in the garden. And along with Death we got a glimpse of ecological collapse and the coming anarchy of an over-peopled planet. And we all, in denial of this unwelcome vision, decided to slam the door and retreat to our toys and make babies—those heirs, those hopes, those products of our species' selfishness, sentimentality, and global death wish.

Cats

I was on a Swan Hellenic educational cruise, #447, which had as its subject the Byzantine Empire, and I was frightfully lacking in comprehension of just about everything. On the second day I went to the ship's library and began reading about Constantinople. I soon became distraught about the fall of the Byzantine Empire. It had happened on a Tuesday. Those fascinating people! That beautiful city! Those awful crusaders, those greedy Venetians.... I read on and on about the triumphant Turks, their baths and harems and eunuchs. They were always drowning people in the Bosphorus or gouging out the eyes of their relatives. There was also much garroting. The palace mutes and dwarves were supposed to be excellent manicurists.... Days followed nights, and I was still engrossed with the Ottomans as our elderly ship *Orpheus* (plucky, but doomed to be scrap the following season) close-cruised Mt. Athos. The cruise had barely begun, and I was already badly off in my timing. Even so, the monasteries, which looked like sprawling failed resorts built against the cliffs, were pretty fascinating and the close-cruising was accompanied by a thrillingly erudite deck talk by one of the lecturers. Women are not allowed on the Holy Island, although it was all the Virgin Mary's idea. Not even female animals are allowed, no goats or chickens or cows or sheep, the reason for this being ... being ... that female animals require too much care or time, time that the monks would more wisely spend in prayer.

After the deck talk a video was shown about Mt. Athos, which of course everyone crowded in to see. I was shocked to see footage of an old monk feeding cats. There must have been twenty-five cats leaping and crawling all over him.

"What about all those cats?" I said. "There has to be a female cat in there somewhere."

"That was explained in the deck talk," an Englishwoman said.

"I missed that," I said. I was embarrassed, but I admitted it.

"An exception was made with the cats," an Englishman said.

Sharks and Suicide

THERE'S SOMETHING OUT THERE WAITING FOR US, AND THAT'S the truth. Wasps or abandoned refrigerators. Dehydration, myxedema, and the three-hundred-year-old elm on the curve. Explosions and wrecks and electrocutions. Funny-tasting meat treats. There are cycles and moments. There are fatal hours. The chop waits in the night and the bright sunshine, and each piece of earth is good enough and greedy for our ending. Dying is the message, all right, but the messengers are bums. Petty. Common. Hasty. We're shot or burned. Our bones start breaking. Cars make waffles of us. We keel over. Or it begins with trouble in the voiding parts. Not many of us die from love or terror these days, and there are few thoughts left that touch us with true horror. But there are some, certainly. There's one. Earth's nightmare is the sea.

Swedenborg said that devils are things that after death choose hell. Who among us knows the extent of the sea's true abyss?

The known food of the West Coast white shark includes pieces of basking sharks (*Cetorhinus maximus*), gray smooth hound sharks (*Mustelus californicus*), Pacific mackerel (*Scomber japonicus*), cabezon (*Scorpaenichthys marmoratus*), halibut (*Paralichthys californicus*), sea otter (*Enhydra lutris*), and man.

The shark as cruising destroyer, acting, for some, as bizarre *machina*, attacks approximately twenty-five people a year. That's hardly many. Nevertheless, the thought of a big fish lunching on a fated bather is known to create concern out of all proportion to the amount of injury or loss of life incurred statistically. Somewhere or other, there's a bronze whaler or a gray nurse or a white death out to do the purely unspeakable. Rolling and trimming, balancing and pivoting, flying with baseball-size eyes through the sea, without malice or immoral intent, a percentage of sharks bite a percentage of people. The chosen can be a silver-suited diver or a black pearler or a little boy in a hemmed T-shirt or a woman bathing with her gentleman friend after lunch or a kid on a fluorescent surfboard with a singing keg.

The pressure exerted by the jaws of a typical eight-foot shark is three metric tons a square centimeter.

The shark is not in the tarot; it is not in the signs of heaven. Its strict reality remains beneath the waters of the world.

A shark attacking a human being probably never strikes from mere hunger, but rather because, under varying circumstances, the victim assumes a suggestion of food, which the shark, out of the purity of its nature, is unable to resist, hungry or not.

The shark likes blood and fish, but it often attacks man when neither of these excitants is present. It strikes in the rain and the bright sunshine, off crowded beaches and in rivers seventy miles from the sea. Neither month nor time of day nor condition of sea or sky nor depth of water nor distance from shore is applicable to the probability of attack. Neither shade of skin nor the presence of sweat or urine is applicable. Nothing is applicable.

The sensory systems of sharks, although beautifully interrelated, nevertheless result in a decidedly limited behavioral repertoire. Once sharks have initiated a specific pattern of behavior, they are not readily distracted or inhibited. Often they continue to attack their prey despite a variety of normally distracting and noxious stimuli, including severe bodily crippling.

Such an object! Both primitive and futuristic with its simple core of mystery, with its actions, exact, obsessed, and inexplicable. Such a deep imperviousness to life! Such silence. Such . . . invisibility. As though created instantly, when needed, out of the sea itself. Day and night, without cessation, death approaches. What can be learned from the shark but negation?

Sharks move boneless from the sea light into the darkness of our worst imaginings. And as the impossible terminus, as the inconceivable hazard, they slip from our dreams into the sea.

SUICIDE

The defining moment for the punk-metal band the Plasmatics was in New York City in the fall of 1980, when Wendy Williams jumped out of a moving Cadillac just before it exploded and catapulted off Pier 82 into the Hudson River. The victim, a '72 Coupe de Ville, had been purchased from a couple who initially had doubts about selling the car they had driven all through their high-school days to the Plasmatics. "I don't want my car to die!" the young wife said.

"Everything must die," Wendy said sensibly, "but your car will be immortalized."

The Plasmatics spent much of the money they got from albums and performances buying the cars and television sets they would then demolish. The pyrotechnical bills were huge. Amplifiers, lighting trusses, guitars—all were subjected to explosive effect. Little could equal the sight of Wendy wailing and shouting, wearing stiletto-heel boots, otherwise accoutered only in shaving cream and a bit of tape over her nipples for modesty's sake, pumping a shotgun loaded with blanks while a false ceiling rigged by her technically inspired crew collapsed around her.

What were they thinking of? Well, it was all about fighting ennui and banality and the torpor of common life, and it could almost remind one of Flaubert, who, writing nostalgically of the friends with whom he shared his youth, said, "We swung between madness and suicide; some of them killed themselves, one strangled himself with his tie, several died in debauchery in order to escape boredom; it was beautiful." Flaubert and his friends, however, were romantics. They didn't say "Fuck You" to the world. Whereas the Plasmatics were all about Fuck You. In their day they were the ultimate in visual anarchy, in staged chaos. "Rock 'n' roll has teeth," Wendy said, in the long-ago '80s. "It will scratch your face off. If you like having your brains blown out and being pushed up against the wall, the Plasmatics are for you."

• • •

Wendy Williams, that's not the same one, right?

This is in Storrs, Connecticut, a dreary state university town, green in the way only the northeast Connecticut landscape can be green, a thick stultifying clogged limp breathless green.

Wendy Williams, that's not the same one, right?

This is what the townsfolk would say when they learned that the thin woman with heavy bangs and long blond braid who rode her bike all around town, going to flea markets and yard sales, buying "blankies" and stuffed toys for the injured and orphaned animals she cared for at the Quiet Corner Wildlife Center was Wendy Williams. Because Wendy Williams was the superhuman screamer with the most fabulous black and blond Mohawk ever, the original hardbody, crazy and kinky and out of control, sexy scary Dada, Wendy Orlean Williams. She had her initials tattooed on her arm! WOW!

Not the same one, right?

Well, she was and she wasn't. She had inhabited her life as a rock-and-roll extremist and speed-metal priestess, and when that life was gone she was just living inside herself, which was Nowhere. The Plasmatics burned inferno bright for four years, from 1978 to 1982, after which Wendy went solo. In 1984 she was on the covers of both *Kerrang!* (the heavy-metal magazine) and *Vegetarian Times*, and in 1985 she was nominated for a Grammy for best female rock vocal. (Tina Turner won.) Then there was the thrash-metal opera *Maggots: The Record* (the fave of Wendy aficionados), which told the tale of a larval takeover of earth and good riddance to it, and then it was 1988 and she was thirty-eight and living in Connecticut.

She didn't mind Connecticut. She had always been capable of disassociative states of consciousness. That's how she functioned so brilliantly as a Plasmatic. This state of being in Connecticut didn't exist for her, Connecticut didn't exist, it was utterly improbable, in no way could it function as a vivarium for Wendy O. Williams. She was used to a lot of time between gigs, but now there was just time. It took her awhile to absorb the fact that her career was over.

In 1993, she tried to kill herself by banging a knife into her chest with a hammer. She changed her mind when the blade lodged firmly in her sternum and she realized that she was afraid that if she pulled it out she'd bleed to death. In 1997 she tried to kill herself again, this time with an overdose of ephedrine, a synthetic version of the health-food store herb *ma huang*—herbal ecstasy. She rejected intervention, of either the pharmaceutical or psychiatric kind. She had never taken drugs. She was a purist in many ways. She had always been outside and against and over the edge in everything, and now, mentally, she had crossed the final threshold Cesare Pavese wrote about in *The Burning Brand*: "It is conceivable to kill oneself so as to count for something in one's own life. Suicide is an act of ambition that can be committed only after one has passed beyond ambition."

Plato said that under certain circumstances, suicide can be justified. These included extraordinary sorrow, unavoidable misfortune, intolerable disgrace. To the Greeks it could be a principled act. But the Christians got hold of it, mixed it up with martyrdom and shortcuts to heaven, and overdid it. By the fifth century, St. Augustine deemed it monstrous, and by the beginning of the fourteenth century, Dante was relegating suicides to the seventh circle of Hell, a mere compartment away from those who had murdered others: There, the souls of suicides are transformed into stunted and twisted trees in a dark and pathless wood. The birds that make their nests there are hideous harpies who tear at the branches and make them cry out and bleed. When Judgment Day comes, these souls will not be reunited with their bodies. The bodies will simply hang from the trees' branches—separate, useless. This grotesque image remains one of the most fearful in all literature. It still moves us, even in our century when few of us believe in Judgment Day and scarcely believe in the soul. We believe in the now and the self. We believe that any anguish we might feel is caused by chemical imbalance, which can be corrected. We believe in the promise of a future through prophylactic drugs. We believe we always have the chance to reinvent ourselves.

Wendy couldn't reinvent herself. She tried to live meekly. She volunteered at the wildlife center. The goal there is to release the injured and orphaned animals back into the wild. You don't look into their eyes; you don't make pets of them. She thought pets were prisoners. She didn't work with dramatic creatures, like raptors. She involved herself almost exclusively with squirrels, which some people regard as decorative rodents. Wendy appeared shy, gentle, soft-spoken, withdrawn. She worked at the Parkade Health Shoppe between Storrs and Hartford. All the employees wore white lab coats like chemists or aides in some sanitorium. Everyone was helpful and cheerful and smiling. It was a normal, utterly preposterous place. Wendy wore the white lab coat and was cheerful and smiling. But there was something spooky about her, too. Her body was in excellent shape, but her mind wasn't—her mind was darting around in dark places, pathless woods. She didn't believe in redemption, but she did believe in deliverance. She was miserable and full of rage, and when she was fired from Parkade and they tried to deny her unemployment benefits, she was deeply pissed. Wendy O. Williams concerned about benefits? It was pathetic. But the benefits weren't the point. It was the whole horrible working in a shopping center, getting your driver's license renewed, getting out of bed in the morning.

So the situation was this: She was living with a tattooist and his cats and suing for unemployment benefits from an employer who didn't want her to have them. Wendy hated cats. She was still taking care of the occasional baby squirrel that had to be nursed with a bottle. She didn't want them to be raised around cats because when they were released they would think cats were their friends, which would make it easy for a cat to mosey up and snap off their heads and disembowel them before they had a chance to say hello. She was working out a couple of hours a week and getting tattooed, and that was all she was doing. The tattoo was too complicated and was making her upset. It was a water and fire design that was not making sense for her. What was it supposed to mean? Innocence and purification? Change in the midst of eternity? Soul mind? What? She had always been physical, an appearance person, and now she hated her body. She had wanted leopard spots, but the boyfriend tattooist couldn't figure out how to do them.

In 1998, April 6 was a Monday. Statistically, April is the most popular month for suicides. Christmas time can't hold a candle to it. And Monday is the favorite day. Wendy took the .38 handgun the tattooist kept beside his bed. What was she thinking? Maybe she was thinking of a happy time, of her birthday, years before, when the Plasmatics gave her a surprise. They were in Phoenix and found out that the city was going to demolish a house to make way for a road. They arranged for a bulldozer, got the proper permits, put her up on that big machine—*Happy Birthday, Wendy!*—and she aimed it and smashed that house to bits.

Of the notes she left, one of them said in part, "I believe strongly that the right to take one's own life is one of the most fundamental rights that anyone in a free society should have." This was the responsibly iconoclastic Wendy, moving from the right to be smutty to the right to die violently by her own hand. One of them said, "My feelings about what I am doing ring loud and clear to an inner ear and a place where there is no self, only calm." This was the dyslexic, philosophic Wendy referring to a place she wanted to be, a place where she wasn't.

Then she went into the woods, fed some squirrels, put a bag over her head so she wouldn't utterly freak out the person who found her, shot herself, and died.

Coral Castle

ED LEEDSKALIN WAS TINY, LATVIAN, AND ODD. LEGEND HAS it he was jilted by his bride-to-be on his wedding day, or maybe the day before, back in 1913, and he came to America, where he wandered forlornly for thirteen years before settling in Florida City. Here he became interested, one might say obsessed, with coral rock, the ancient dead reef that makes up the foundation of Florida's Upper Keys. The rock had been quarried extensively for bridge and causeway fill for Henry Flagler's railway to Key West, and there was a lot of it lying around, very heavy of course, weighing many tons. Using chains, pulleys, and old car parts, Ed moved chosen pieces of rock—which is really beautiful in its polyp-and-gorgonian-riddled way—and began fashioning it into tables and chairs and planets and even a moon, a large crescent moon weighing twenty-three tons. Then he moved, taking all his rock with him. He bought some land in Homestead, ten miles away, and transported everything in a borrowed truck, alone, in the dead of night, legend has it.

In Homestead he really went to work. For twenty years! He cut and carved coral to make ponds and settees, steps going up to nothing and steps going down into the ground, towers and telescopes, a sundial. He made tons of chairs, a heart-shaped "Feast of Love" table, a rock sun-couch, and a bedroom full of slabby beds, large and small, also a rocking cradle. He thought it would be nice to have children, and he looked forward to disciplining them in the "Repentance Corner." If they were naughty, he would make them put their little heads in holes he cut in the rock. Their heads would be held in place with a wedge of wood, and he would lecture those little heads.

When he was more or less finished with his work, he built a wall around it all, a lovely eight-foot-tall wall of perfectly fitted coral blocks and the pièce de résistance, a nine-ton gate so perfectly balanced it swung open with the touch of a finger. He lived alone behind the wall in a tiny room, on a bed made of boards wrapped in burlap and hung from the ceiling. He never had a girl. He wrote in his journal, "I ALWAYS WANTED A GIRL BUT I NEVER HAD ONE," not even the girl who jilted him at the altar, the sixteen-year-old by the name, legend has it, of Agnes Scuffs. Agnes Scuffs of Latvia.

It's just as well Agnes didn't marry her suitor. Her life would have been so hard. She would have eaten only green vegetables and wild rabbits. She would have been up all night fiddling with that coral rock. Leedskalin's idea of fun was closing his eyelids and giving a side-push to the eyeball to see tiny bolts of lightning dancing about. Maybe Agnes wouldn't have thought it was that much fun. She might have wearied of his endless theorizing and philosophizing. Then there were the magnets. She would have had to live with those adored magnets. Magnets were everything to him. "SURPLUS MAGNETS ARE THE REAL LIFE," he wrote. Magnets were "THE COSMIC FORCE. THEY HOLD TOGETHER THIS EARTH AND EVERYTHING ON IT AND THEY

HOLD TOGETHER THE MOON TOO." The only thing keeping the moon up was magnets. To bring the moon down you'd just give it a half turn so it was no longer aligned with its magnets. If Agnes disagreed, he would have said, as he did to everyone, "YOU SAY I'M WRONG, I SAY YOU ARE WRONG YOURSELF." He believed the body got its magnets from food; digestive acids dissolved the food and liberated magnets to be used for other purposes (in his case, moving thousands of tons of rock all by himself). Maybe it's best the little fellow's love remained unrequited. Everyone knows unrequited love is the strongest magnetic force of all.

One of the nicest things to happen to Coral Castle was in 1963, when the fabulously unclassifiable movie director Doris Wishman shot one of her early nudie films there, *Nude on the Moon*. Watching *Nude on the Moon* is a fine way to visit the castle. In Doris's film, spacemen land on the moon and discover that nudists are that planet's sole inhabitants. The naked actors frolicked amidst the rocks, quite at home among the throne and obelisks and the eighteen-ton models of Mars and Saturn. The nudists conferred at a coral table shaped like Florida, while communicating telepathically through little quivery antennae on their heads. This method enabled Doris to dub all the sound in afterwards, the nudists not being terrifically competent actors.

Ed Leedskalin himself may have come from the moon or somewhere farther. When he talked about "sweet sixteen," which he did a lot, everyone presumed he was talking about Agnes Scuffs of Latvia, but in fact he was not. *Sweet Sixteen* was the spaceship he thought might find the Coral Castle and take him home, back home to the Mother Rock somewhere in space. But the spaceship never found the castle, and when he was sixty-four, he checked himself into a Miami hospital and died. Some say it was malnutrition, others leukemia. In either case his magnets failed him. Maybe it was because he had stopped moving those rocks around, carving out the chairs so soft and comfy, changing that coral rock's hard, hard nature, hauling and shaping all those mutely calling-out things. Maybe the rocks didn't like him stopping.

There's a Pizza Hut across the street from Coral Castle now. It might have the most direct harmonic relationship with magnetic forces of any Pizza Hut in the world. Or it might not. No one eating there has mastered levitation. And no one has mentioned seeing tiny Ed either, five feet tall and less than one hundred pounds, wearing his dark suit and tie. But why would they? He wanted to be cremated so that he would eventually turn into a rock. He had no desire to return as Ed Leedskalin. They say his secrets died with him. As will ours all.

One Acre

I HAD AN ACRE IN FLORIDA ON A LAGOON CLOSE BY THE GULF of Mexico.

I am admittedly putting this first line up against Dinesen's famous oneiric one: *I had a farm in Africa at the foot of the Ngong Hills.* When Dinesen first came to Africa, she confessed that she could not "live" without getting a fine specimen of every single kind of African game. The hunt, for her, was an eroticized image of desire, "a love affair" wherein the "shot was . . . in reality . . . a declaration of love." She must have blushed to read this drivel later, for after ten years she found hunting "an unreasonable thing, indeed in itself ugly and vulgar, for the sake of a few hours enjoyment to put out a life that belonged in the great landscape and had grown up on it." One could say her thinking had evolved, that she had become more conscientious. Still, when she was about to leave her beloved farm (her house, empty of furniture was admirably "clean like a skull"), she planned to shoot her dogs and horses, dissuaded from doing so only by the pleas of her friends. The animals belonged to her, as had the land, which she ceased to own only when it became owned by another and subject to that person's whims and policies. Of course, it became hers again through the writing about it.

She preserved it in *Out of Africa*. Once again, Art, reflective poesy, saves landscape.

I had an acre in Florida. . . . This bodes no drama. For what wonders could a single acre hold, what meaning or relevance? Though the word *Florida* is oneiric too, and thus its own metaphor. It is an occasional place, a palmed and pleasant stage for transients. To hold fast to an acre in that vast state is almost neurotic. An acre is both too much and not enough. Its value lies in its divisibility. How many building lots are permitted by law? Four, certainly.

I once saw a white heron on the sprawling outskirts of Naples, a city that crowds against the Big Cypress Preserve and Everglades National Park. The heron seemed to be beating its head against a tree knocked down by bulldozers to widen a road. Water still lay along the palmetto-dotted earth, but pipes would soon carry it away and dry the land for townhouses and golf courses. Cars sped past. The heron, white as robed angels must surely be, was beating his head against the tree. He was lost to himself, deranged, in his ruined and lost landscape.

I have seen numbers of water birds struck down by cars. I used to take them home and bury them between the mangroves and the live oaks on my lagoon, though of course it was not my lagoon. It is only a mile and a half long. To the north, it cedes to a private road that gives access to the Sanderling Club, where the exceptionally wealthy enjoy their gulf views. To the south it vanishes beneath the parking lot for a public beach. This is on Siesta Key, a crowded seven-mile-long island off Sarasota that is joined to the mainland by two bridges, one of four lanes, the other two. The lagoon is named Heron, the beach, Turtle. Yes, the turtles still come to nest, and the volunteers who stake and guard the nests are grateful—they practically weep with gratitude—when the condo dwellers keep their lights out during the hatching weeks so as not to confuse the infant turtles in their night search for the softly luminous sea. But usually the condo dwellers don't keep their lights out. They might accommodate the request were they there, but they are seldom there. The lights are controlled by timers and burn bright and long. The condos are investments, mostly, not homes. Like the lands they've consumed, they're cold commodities. When land is developed, it ceases being land. It becomes covered, sealed, its own grave.

Ecosystems are something large, to be saved, if at all, by the government at great expense and set aside to be enjoyed by all of us in some recreational or contemplative fashion. An individual doesn't think of himself as owning an ecosystem. The responsibility! Too much. Besides, there's something about the word that denotes the impossibility of ownership. *Land*, on the other hand, is like a car or house; it has economic currency. It is a marketable concept, far from Aldo Leopold's definition of it as *a fountain of energy flowing through a circuit of soils plants and animals*. It was over fifty years ago that Leopold wrote his elegantly reasoned essay "The Land Ethic," but it has had about as much effect on the American conscience as a snowflake. Seven thousand acres are lost each day in this country to development. Ecosystem becomes land becomes parcel. The fountain of energy is shut down.

On Siesta Key, so-called "open space" is being utilized by the county as beach parking or public tennis courts. "Raw" land no longer exists, though a few lots are still available, some with very nice trees, most of which will have to go (unfortunately) in order to accommodate the house that will be built on what is now considered a "site." Hardly can get all ecosystem emotional over a site. Should a banyan tree be growing there, it will most assuredly have to go because it is in their nature to grow extravagantly and demand a great deal of space. Trees, of course, cannot demand anything. Like the wild animals who have certain requirements or preferences—a clutter and cover, long natural hours of friendly, concealing dark—anything they need can be ignored or removed right along with them.

In 1969, I bought Lots 27, 28, and 29 on Midnight Pass Road, a two-lane road that ended when the key did. There was a small cypress house, no beauty, and an even smaller cypress cottage. They were flat-roofed single-story affairs, built on poured-concrete slabs. The lots together cost $24,000. In 1972, I bought Lots 30 and 31 for $12,000. Lagoon land wasn't all that desirable. There was no access to open water. Bay land was more valuable, and, even then, gulf front, with its inherently "protected" view, was only for the wealthy. The view is "protected." I could hear the gulf from my small acre; it was in fact only several hundred feet away, concealed by a scrim of mangroves. The houses that were to be built over there were grand but still never quite exceeded the height of the mangroves. I did not see my lagoon neighbors for my trees, my tangled careless land, though as the years passed, I put up sections of wooden fence, of shadowbox design, for my neighbors changed and multiplied. The little cypress houses so similar to mine were torn down in a twinkling, the "extra" lots sold. I put a wooden fence up along the road eventually. It weathered prettily but would shudder on its posts from a flung beer bottle, and sections of it were periodically demolished by errant cars. I don't believe I ever rushed sympathetically to the befuddled driver's aid. Streetlights went up at fifty-foot intervals on the dark and curvy road. The bay side got the lights, the lagoon side got the bicycle path. Home owners were responsible for keeping the "path" tidy, and I appeared out there dutifully with broom and rake, pushing away the small oak leaves from the trees that towered overhead, disclosing all that efficient concrete for the benefit of increasing streams of walkers and joggers. Bicyclists preferred to use the road. Any stubborn palmetto that fanned outward or seeded palm that once graced the strip of land outside my rickety wall would be snipped back by a supernumerary, doing his/her part for the public way. The bottles, cans, and wee chip bags were left for me to reap. As owner of Lots 27, 28, 29, 30, and 31, I had 370 feet of path to maintain. I became aware, outside my fence, of the well-known Florida light, a sort of blandly insistent urban light—feathery and bemused—not insistent but resigned. Cars sped past. Large houses were being constructed on the bay, estates on half an acre with elaborate wrought-iron fences and electric gates. Palmetto scrub had given way to lawns. Trees existed as dramatically trimmed accents, all dead wood removed. Trees not deemed perfectly sound by landscape professionals were felled, the palms favored were "specimen" ones. Dead animals and birds appeared more and more frequently on the road. The cars sped past incessantly.

Within my wobbly fence, I pottered about. The houses were built in the 1940s, and the land had the typical homesteaded accoutrements of that time—a few citrus trees, some oleander and hibiscus for color, a plot cleared for a few vegetables and shasta daisies, a fig tree left to flourish for shade, and live oaks left to grow around the edges. The ghastly melaleucas were available in nurseries then and were often planted in numbers as a hedge. The man I bought the land from was a retired botanist, and he had planted avocado and lychee nut trees, too, as well as a grove of giant bamboo from which he liked to make vases and bowls and various trinkets. There was bougainvillea, azalea, gardenia, powder puff, and firecracker plant, crotons, wild lilies, sea grape, and several orchid trees. Of the palms there was a royal, sabal, many cabbage, pineapple, sago—queen and king—reclanata, fishtail, sentry, traveler's, and queen's. There were cypresses, a jacaranda, and two banyan trees. There was even a tiny lawn with small cement squares to place the lawn chairs on. The mangroves in this spot had been cut back for a view of the idly flowing tea-colored lagoon. Elsewhere, they grew—the red and the black—in the manner each found lovely, in hoops and stands, creating bowers and thickets and mazes of rocking water and dappling light.

This was my acre in Florida. Visitors ventured that it looked as though it would require an awful lot of maintenance, though they admired my prescience in buying the extra lots, which would surely be worth something someday. The house had a certain "rustic charm," but most people didn't find the un-air-conditioned, un-dehumidifed air all that wholesome and wondered why I kept the place so damn dark, for there were colored floodlights widely available that would dramatize the "plantings." I could bounce more lights off the water; you could hardly even tell there was water out there, and what was the sense of hiding that? And despite the extraordinary variety, my land seemed unkempt. There were vines and brazilian pepper and carrotwood, there were fire ant mounds, rats surely lived in the fronds of the untrimmed palms. My acre looked a little hesitant, small and vulnerable, young. Even the banyan tree was relatively young. It had put down a few aerials but then stopped for a good decade as if it were thinking . . . *What's the use? I'm straddling Lots 29 and 30 and I'm not known as an accommodating tree. When the land gets sold, I'll be sold too and will be felled in screaming suttee.* . . . Or sentiments of that sort.

As for the birds and animals, well, people didn't want raccoons and opossums and armadillos, and their cats would eat the baby rabbits. Too disgusting, but that's just the way nature was. And though I had cardinals and towhees and thrashers and mockingbirds and doves and woodpeckers, they did too; as a matter of fact, their cardinals were nesting in a place where they could actually see them, right near the front door, although that was getting to be a nuisance. As for the herons, you found them everywhere, even atop the Dumpsters behind the 7-Eleven. They were flagrant beggars. You had to chase them away from your bait bucket when you were fishing from the beach. Did I fish in the lagoon? There were snapper in there, redfish, maybe even snook. I could get the mullet with nets. Why didn't I fish?

The years flowed by. Some of the properties on the lagoon fell to pure speculation. Mangroves were pruned like any hedging material; in some cases, decks were built over them, causing them to die, although they remained ghostily rooted.

Landowners on the gulf did not molest their mangroves. The lagoon to them was the equivalent of a back alley, why would they want to regard the increasing myriad of houses huddled there? I traveled, I rented the place out, I returned. There were freezes, we were grazed by hurricanes. An immense mahoe hibiscus died back in a cold snap, and two years later, a tall, slender, smooth-barked tree it had been concealing began producing hundreds of the pinkest, sweetest, juiciest grapefruit I have ever tasted. The water oaks that had reached their twenty-year limit rotted and fell. There were lovely woodpeckers. All through the winter in the nights the chuck-will's-widow would call.

That would drive me nuts, several of my acquaintances remarked.

The sound of construction was almost constant, but no one appeared to be actually living in the remodeled, enlarged, upgraded properties around me. I had cut out sections of my side fences to allow oak limbs to grow in their tortured specific manner, but my neighbor's yard men would eventually be instructed to lop them off at the property line. This was, of course, the owner's right. There was the sound of trimmers, leaf blowers, pool pumps, pressure cleaners, the smell of chemicals from pest and lawn services. Maintenance maintenance maintenance. Then the county began cutting back the live oak limbs that extended over the bicycle path even though one would have to be traversing on a pogo stick to bump into one. Sliced sure as bread, the limbs, one at least five feet in diameter and green with resurrection fern and air plants, were cut back to the fence line.

It was then I decided to build the wall.

The year was 1990. The wall was of cement block with deep footers, and it ran the entire length of the road footage except for a twelve-foot opening that was gated in cedar. It cost about $10,000, and two men did it in two months. The wall was ten feet high and not stuccoed. I thought it was splendid. I didn't know many people in the neighborhood by then, but word got back to me that some did not find it attractive. What did I have back there, a prison? To me, it was the people speeding past the baby Taj's on the animal-corpse-littered road who had become imprisoned. Inside was land—a mysterious, messy fountain of energy—outside was something else, not land in any meaningful sense but a diced bright salad of colorful real estate, pods of investment, its value now shrilly sterilely economic.

Behind the wall was an Edenic acre, still known to the tax collector as Lots 27, 28, 29, 30, and 31. Untransformed by me, who was neither gardener nor crafty ecological restorer, the land found its own rich dynamic. Behind the wall was neither grounds nor yard nor garden or park or even false jungle but a functioning wild landscape that became more remarkable each year. Of course there was the humble house and even humbler cottage, which appeared less and less important to me in the large order of things. They were shelters, pleasant enough but primarily places from which to look out at the beauty of a world to which I was irrelevant except for my role of preserving it, a world I could be integrated with only to the extent of my not harming it. The wildlife could hardly know that their world in that place existed only because I rather than another *owned* it. I knew, though, and the irrationality of the arrangement, the premise, angered me and made me feel powerless, for I did not feel that the land was mine at all but rather belonged to something larger that was being threatened by something absurdly small, the ill works and delusions of—as William Burroughs liked to say—homo sap.

Though the wall did not receive social approbation, its approval from an ecological point of view was resounding. The banyan, as though reassured by the audacious wall, flung down dozens of aerial roots. The understory flourished, the oaks soared, creating a great, grave canopy. Plantings that had seemed tentative when I bought them from botanical gardens years before took hold. The leaves and bark crumble built up, the ferns spread. It was odd. I fancied that I had made an inside for the outside to be safe in. From within, the wall vanished; green growth pressed against it, staining it naturally brown and green and black. It muffled the sound and heat of the road. Inside was cool and dappled, hymned with birdsong. There were owls and wood ducks. An osprey roosted each night in a casuarina that leaned out over the lagoon, a tree of no good reputation and half dead, but the osprey deeply favored it, folding himself into it invisibly each nightfall. A pair of yellow-crowned night herons nested in a slash pine in the center of Lot 30. Large birds with a large hidden nest, their young—each year three!—not hasty in their departure. A single acre was able to nurture so many lives, including mine. Its existence gave me great happiness.

And yet it was all an illusion too, its own shadowbox, for when I opened the gates or canoed the lagoon, I saw an utterly different world. This was a world that had fallen only in part to consortiums of developers; it had fallen mostly plat by plat to individuals, who, paradoxically, were quite conformist in their attitude toward land, or rather the scraped scaffolding upon which their real property was built. They lived in penury of a very special sort, but that was only my opinion. In their opinion they were living in perfect accord with the values of the time, successfully and cleverly, taking advantage of their advantages. Their attitudes were perfectly acceptable, they were not behaving unwisely or without foresight. They had maximized profits and if little of nature had been preserved in the arrangement, well nature was an adornment not to everyone's taste, a matter really of personal tolerance and sympathy. Besides, Nature was not far away, supported by everyone's tax dollars and preserved in state and federal parks. And one could show one's appreciation for these places by visiting them at any time. Public lands can be projected as having as many recreational, aesthetic, or environmental benefits as can be devised for them, but private land, on this skinny Florida key and almost everywhere in this country, is considered too economically valuable to be conserved. Despoilation of land in its many many guises is the custom of the country. Privately, one by one, the landowner makes decisions that render land, in any other than financial terms, moot. Land is something to be "built out."

In contrast to its surrounds, my acre appeared an evasion of reality, a construct, a moment poised before an inevitable after. How lovely it was, how fortunate I was. Each day my heart recognized its great worth. It was invaluable to me. The moment came when I had to sell it.

Leopold speaks of the necessity of developing an ecological conscience, of having an awareness of land in a philosophical rather than an economic sense. His articulation of our ethical obligations to the land is considered by many to be quite admirable. We celebrated the fiftieth anniversary of this articulation (if not its implementation) in 1999. A pretty thought, high-minded. And yet when one has to *move on* (if not exactly in the final sense), one is expected to be sensible, realistic, even canny, about property. I was not in the comfortable kind of financial situation where I could deed my land to a conservation group or land trust. Even if I could have, it would probably have been sold to protect more considerable sanctuary acreage elsewhere, for it was a mere acre in a pricey neighborhood, not contiguous with additional habitat land, though the lagoon did provide a natural larger dimension. I had been developing an ecological conscience for thirty years, and I could continue to develop it still, certainly become a good steward somewhere else, because once I had decided to sell, this particular piece of land and all the creatures that found it to be a perfect earthly home, would be subject to erasure in any meaningful ecological sense, and this would not be considered by society to be selfish, cruel, or irresponsible.

"Wow, it's great back here," the Realtor said. "I often wondered what the heck was going on back here. I'm looking forward to showing this place."

I told him I wanted to sell the land as a single piece, with deed restrictions, these being that the land could never be subdivided; that the buildings would be restricted to one house and cottage taking up no more land than the originals; and that the southern half of the property would be left in its natural state as wildlife habitat.

"Nobody wants to be told what they can do with their land." He frowned. "I'll mention your wishes, but you'll have to accept a significant reduction in price with those kinds of demands. When we get an offer, you and the buyer can negotiate the wording of the agreement. I'm sure the type of person who would be attracted to this property wouldn't want to tear it all apart."

"Really?" I said. "You don't think?"

I went through a number of Realtors.

With a lawyer I drew up a simple and enforceable document that the Realtors found so unnerving that they wouldn't show it right away to interested parties, preferring word wobble and expressions of good intention. There were many people who *loved* the land, who *loved* nature but would never buy anything that was in essence not free and clear. Or, they had no problem with the restrictions *personally*, but when they had to sell (and heaven forbid that they would right away, of course), they could not impose such coercive restraints on others. The speculators and builders had been dismissed from the beginning. Those interested were people of a more maverick bent, caring people who loved Florida, loved the key—wasn't it a shame there was so much development, so much change. When they saw the humble document, they said 1) who does she think she is, 2) she's crazy, 3) she'll never sell it. Over the months the Realtors took on a counseling manner with me, as though I needed psychological guidance through this problem of my own making, as though I needed to be talked down from my irrational fanciful resolve. They could sell the land for $200,000 more if I dropped the restrictions. They argued that my acre could be destroyed naturally, a hurricane could level everything and the creatures, the birds, would have to go somewhere else anyway. With the money I'd make marketing it smartly, I could buy a hundred acres, maybe more, east of the interstate. There was a lot of pretty ranchland over there. I could conserve that. A lot of pressure would be on that land in a few years, I could do more by saving that. Sell and don't look back! That's what people did. You can't look back.

I'm not looking back, I said.

And I wasn't.

I was looking ahead, seeing the land behind the wall still existing, still supporting its nests and burrows—a living whole. I was leaving it—soon I would no longer be personally experiencing its loveliness—but I would not abandon it. I would despise myself if I did. If I were to be party to a normal real estate transaction, I would be dooming it; I would be—and this is not at all exaggerated—signing a warrant for its death. (Perhaps the owners of the four new houses that would most likely be built would have the kindness to put out some bird seed.) I wanted more than more money for my land, more than the mere memory of it, the luxury of conserving it falsely and sentimentally through lyrical recall. I wanted it to be.

It took eight months to find the right buyer. Leopold's "philosophers" were in short supply in the world of Florida real estate. But the ideal new owners eventually appeared, and they had no problem with the contract between themselves and the land. I had changed no hearts or minds by my attitude or actions, I had simply found—or my baffled but determined Realtors had—people of my persuasion, people who had a land ethic too. Their duties as stewards were not onerous to them. They did not consider the additional legal documents they were obliged to sign an insult to their personal freedom. They were aware that the principle was hardly radical. An aunt had done a similar thing in New England, preserving forty acres of meadow and woodland by conservation easement. They had friends in California who had similarly sold and conserved by deed four hundred acres of high desert. And here was this enchanted acre.

It had been accomplished. I had persisted. I was well pleased with myself. Selfishly, I had affected the land beyond my tenure. I had gotten my way.

And with all of this, I am still allowed to miss it so.

Audubon

THE ROSEATE SPOONBILL HAD ALMOST BEEN EXTERMINATED
BY plume hunters at the turn of the century. The feathers were
not as popular as those of the egret for ladies' hats, but the
wings were torn off and made into fans, though the buyer was
often disappointed when the brilliant colors quickly faded.
Knowledge of the horrors of avian carnage gives reading Edith
Wharton a new dimension.

The spoonbill is a simple and shy creature of many trou-
bles, yet garbed in glory. The young are an immaculate white
and only gradually become suffused with pink. Three years
must pass and three moltings occur before the bird achieves
its full brilliance of nuptial plumage—in rose and carmine and
orange—and will mate. The drawing of the spoonbill in John
Audubon's *Birds of America*, considered not to be by Audubon, does not reflect the true radiance of the bird's colors. It is
not likely that Audubon saw many of them. In his remarks he
noted that their flesh was oily and poor eating, and that they
were difficult to kill.

When Audubon visited Key West in 1832, a newspaper
editor wrote enthusiastically:

"It is impossible to associate with him without catch-
ing some portion of his spirit; he is surrounded with an

atmosphere which infects all who come within it, with a mania for bird killing and bird stuffing."

Audubon indeed had a mania. Though his name has become synonymous with wildlife preservation, he was in no manner at any time concerned with conservation. He killed tirelessly for sport and amusement as well as for his art, and he considered it to be a very poor day's hunting in Florida if he shot fewer than a hundred birds. From St. Augustine, he wrote: "We have drawn seventeen species since our arrival in Florida but the species are now exhausted and therefore I will push off. . . ."

Audubon shot many thousands of birds and never in his mind made the connection between the wholesale slaughter he so earnestly engaged in and their decreasing number, although in his forties he did begin complaining about the scarcity of mammals and birds for his studies. "Where can I go now," he grumbled, "and visit nature undisturbed?"

The Animal People

FOR CENTURIES POETS, SOME POETS, HAVE TRIED TO GIVE
A voice to the animals, and readers, some readers, have felt
empathy and sorrow. If animals did have voices, and they
could speak with the tongues of angels—at the very least with
the tongues of angels—they would be unable to save themselves from us. What good would language do? Their mysterious otherness has not saved them, nor have their beautiful
songs and coats and skins and shells and eyes. We discover the
remarkable intelligence of the whale, the wolf, the elephant—
it does not save them, nor does our awareness of the complex-
ity of their lives. Their strength, their skills, their swiftness,
the beauty of their flights. It matters not, it seems, whether
they are large or small, proud or shy, docile or fierce, wild
or domesticated, whether they nurse their young or brood
patiently on eggs. If they eat meat, we decry their viciousness;
if they eat grasses and seeds, we dismiss them as weak. There is
not one of them, not even the songbird who cannot, who does
not, conflict with man and his perceived needs and desires.
St. Francis converted the wolf of Gubbio to reason, but he
performed this miracle only once and as miracles go, it didn't
seem to capture the public's fancy. Humans don't want animals to reason with them. It would be a disturbing, unnerving,

diminishing experience; it would bring about all manner of awkwardness and guilt.

We learn more and more about them, and that has not saved them. We know that when they face Death, they fear it. We know that they care for their young and teach them, that they play and grieve, that they have memories and a sense of the future, for which they sometimes plan. We know about their habits, their migrations, that they have a sense of *Home*, of finding, seeking, returning to Home. We know these things, and it has not saved them. *We know where they live* on this planet, and nine times out of ten we will go there and . . . rout . . . them . . . out. Nothing that is animal, that is not us, cannot be slaughtered as a pest or sucked dry as a memento or reduced to a trophy or rendered into a product or eaten, eaten, eaten. The French eat horses, the Japanese whales, the Taiwanese dogs. Gorillas and chimpanzees are now being killed in quantity on a commercial level, to provide bushmeat for African workers who are decimating their forests for European timber companies. It's need or preference or availability, it's culture, it's a way to feed the poor, it's different, it's plentiful, it's not plentiful but it's a nice change of pace, it arouses the palate, it amuses the palate, it's healthy, it gets rid of something unwanted, it utilizes what's already dead, it would live too long otherwise and take up too much space, it's somebody's way of life, it's somebody's livelihood, it's somebody's business, it's an industry.

The creatures that have been under our "stewardship" the longest, who have been codified by habit for our use, the farm animals, have never been as cruelly kept or confined or slaughtered in such numbers in all of history. They have always suffered a special place in our regard—they are *known* to us, they are tamed, they are *raised* to provide us with milk and eggs and meat, they are bred to die. Large-scale corporate agribusinesses are pure Descartes. Animals are no more than machines—milk machines, piglet-making machines, egg-laying machines. Production units converting themselves into profits. Pigs are raised on bare concrete, in windowless metal buildings, or tightly restrained in foul pens and gestation boxes. Two hundred and fifty thousand laying hens can be confined in a single building on a factory farm with fully automated egg collection. The high mortality rate caused by overcrowding is considered economically acceptable. Nothing is more worthless than an individual chicken. Cows are kept pregnant to produce milk, the amount of which is artificially increased with synthetic hormone injections, although the dairy industry already produces enormous quantities of excess milk. The by-products of the dairy industry, calves, are chained in crates twenty-two inches wide and no longer than their bodies and raised on a diet of drug-laced liquid feed for a few months until they're slaughtered for the delicacy veal. The factory farm today is a crowded, stinking bedlam, filled with suffering animals who are quite literally insane, sprayed with pesticides and fattened on a diet of growth stimulants, antibiotics, and drugs, some of which, like sulfamethazine and clenbuterol, are proven carcinogenics.

Sliced, cubed, shrink-wrapped, their remains in our vast, spotless supermarkets have borne no resemblance to living things in our minds for some time now—they are merely some things, in another department from the vegetables. The super-market is not a place where one thinks . . . Animal. Now, even at the source of their lives, where they live their brief lives, they are, in our time, slipping away from being thought of as animals at all. They are explicitly excluded from any protection offered by the Federal Animal Welfare Act, an act that is casu-ally and lightly enforced, if at all, by the Department of Agri-culture. "Normal agricultural operation" precludes humane treatment and anticruelty laws do not apply to that which is raised for food. I thought that no one ate veal anymore. After learning how they were raised in darkness, in crates, not in a meadow on mother's milk at all, I just assumed. . . . But, no, many people say, well, apparently they're raised in the dark, in crates or something, but the taste is creamy, refined, I like it. . . . Gourmands will stop eating veal only if they become convinced that they'll get a killer disease if they don't.

In England, the beef industry had a setback when a link was found between bovine spongiform encephalopathy (BSE), a fatal disease of cattle, and Creutzfeldt-Jakob disease, a fatal neurological virus in humans. The cows became ill because they were fed the rendered remains of sick sheep. Of course, in this country we are assured that our cows aren't being fed sick sheep and that no BSE-infected cattle have been found here. We do have many "downer" animals, though, about 100,000 of them a year, that collapse from stress or something, heaven knows, and end up dead prior to the slaughtering process. They are rendered and ground up and become pet food and animal feed. Cattle do eat cattle here. They are fed the ground offal of those that have succumbed to unknown causes, and this has been the practice for many years. If BSE were ever confirmed in this country, which is not at all unlikely, people would stop eating meat for a while for the same reasons the English did. Not because they'd had a sudden telepathic vision of the horrors of the abattoir, or because they'd all been subjected to a reading of James Agee's "A Mother's Tale" but because they thought that eating steak would make their brains go funny. It's unlikely that they'd turn to vegetarianism.

Vegetarians are still regarded somewhat suspiciously and, in general, not admired much. Their Meat is Murder chirping seems to be an irritant right up there with the noise of a leaf blower or a Jet Ski. And their wishful hope that by their example, animals will be saved, that slaughterhouses will fall silent and "modern" agribusiness will crumble, seems naive, for elementary economics do not apply to the world of agribusiness. If the state of Maine, say, went vegetarian or the entire state of Florida, led by those tanned oldsters who have finally for the first time, grasped what Ecclesiastes 3:19 is saying—went all the way vegan, what would happen?

Well, such a fundamental, *abnormal* shift in attitudes would constitute a crisis for many thousands of people in the intensive agricultural industry. On an average day in America, 130,000 cattle, 7,000 calves, 360,000 pigs, and 24 million chickens are killed, and you can't shut a show like that down overnight. On the nightly news there'd be footage of cows and calves being shot, pigs being bulldozed into pits, chickens being gassed and dumped—for *nothing*. Vegetarians would be accused of causing the carnage—the blood would be on their hands because they hadn't been realistic, they hadn't thought their actions through, they hadn't realized how illogical and egotistical they were being. Too, the argument goes, even when people don't eat animals (unless they are zealots or Jains), they are culpable in their deaths. For as well as being turned into the more obvious sofas, shoes, and jackets, animals are transmogrified into anti-aging creams and glue and paint and cement and condoms. Gelatin—benign gelatin, formerly known as hooves—constitutes Jell-O, of course, and is also in ice cream and the increasing number of fat-free products we consume. Animals are turned into all manner of drugs, mood enhancers, and mood stabilizers. Premarin, an estrogen drug for menopausal women, comes from the urine of pregnant mares. This is a whole new industry that results in the births of approximately 75,000 unwanted foals each year. Off to the slaughterhouse the little ones go, along with the big racehorses who had the misfortune never to see a winner's circle. Prozac was developed in the laboratory by—well, to make a long story short—injecting rats with various compounds and blenderizing their brains, which were then injected into other rats whose brains were blenderized, and so on to discover a chemical that would block a postulated cause of depression: too many neuro-serotonin particles. Not that rat brains are Prozac *per se,* but it could be said that millions of dead rats are responsible for our being so at peace with ourselves.

Animals are everywhere in our lives. (We just can't look into their eyes. And we've gotten used to not looking into their eyes.) We distance ourselves from them more and more as we use them in increasingly unnatural ways. They're practically on the verge of being reclassified, so that our remaining compassion and ethical concerns for them will be made irrelevant. In the laboratory, animals are tools, they're part of the scientific apparatus, they undergo transformation, they are metamorphized into data. Rats and mice are already excluded from the definition of *animal* by the Department of Agriculture. Rats and mice are simply not animals, they are something else. And to take it one step further these un-animals are then genetically manipulated and reinvented. Hairless mice were created some time ago to make it easier for researchers to administer injections. There are countless variations of mutant "knockout" mice that lack particular genes crucial to learning or instinctual behavior. There are AIDS mice and cancer mice who self-destruct in novel ways. There are countless creatures in the labs whose genetic code had been permanently altered, creatures programmed to suffer (though suffering to experimenters is considered a theoretical abstraction); to be born with or develop terrible diseases and deformities. The first patent for a genetically altered animal was granted in 1987. The engineering, altering, and manufacturing of animals has barely begun. A side benefit of this is that we don't have to feel guilty about "animals" anymore. Any sentience they'd possess would be invented by man or could be eliminated altogether. Animals would have no more a real "life" than a lightbulb.

Animals of the farm, manipulated through drugs to grow faster and larger, to produce more milk or leaner meat, are commonplace. Now there is "pharming," a logical continuance of this accepted trend. Researchers are creating entire new orders of creatures—specifically designed, transgenic, xenograph-ready. Around the world in labs with names such as Genpharm

International Inc., Genzyme Corporation, and Pharmaceutical Proteins, biotechnocrats are inserting human genes into livestock to form animals that can produce human proteins and hormones: drugstores on the hoof. Pigs, long attractive to the farmer, not because of any Babe or Miss Piggy–like charm but because they have short pregnancies and big litters, have become a favorite of researchers, who are altering them to make the perfect organ donors. Humans are requiring fresh new organs all the time, and employing animals in this way seems so much more sophisticated than merely eating them. The ethics of breeding animals for body parts to replace our own failing ones seems to give people pause only when combined with warnings of dangers to human health. A person might not want that little monkey's heart, not because he wanted the monkey to keep it but because he'd worry that he might contract the Ebola virus and his skin would get pulpy, he'd vomit black blood, and his eyeballs would burst.

If, however, you found this fear beside the point—if, in fact, you felt that the monkey had a right to his own heart— you would be considered somewhat of an oddball. You'd be one of the "animal people." You'd believe in animal liberation, you would be part of the animal rights movement. Technology presses to remove animals from nature, to muddy and morph the remaining integrity of the animal kingdom. Technocrats would further reduce animals, historically considered property under the law, to defined, even designed varieties of use. While the animal rights movement is attempting to pierce the barrier between species in order to give animals equal consideration under the law, far more powerful economic forces want to pierce the barrier for far different reasons and to far different effect. While the animal rights movement tries to make people face their responsibilities to the living world and question prevailing nature-breaking values, technology merely dismisses their concerns and characterizes the movement as being composed of crazies and cranks.

Anthropomorphism originally meant the attribution of human characteristics to God. It is curious that the word is now used almost exclusively to ascribe human characteristics—such as fidelity or altruism or pride, or emotions such as love, embarrassment, or sadness—to the nonhuman animal. One is guilty of anthropomorphism, though it is no longer a sacrilegious word. It's a derogatory, dismissive one that connotes a sort of rampant sentimentality. It's just another word in the arsenal of the many words used to attack the animal rights movement.

• • •

The American Medical Association, the National Association for Biomedical Research, the Food and Drug Administration, the National Institutes of Health, and the biotechnological, chemical, and pharmaceutical companies of this our land have an idea of what a day in the life of an animal rights activist is like, and they want to share it with you:

Wearing aggressive T-shirts that quote the maunderings of people you've never heard of—Schweitzer, Schopenhauer, Thomas à Kempis—animal activists start their morning by participating in one or another annoying obstructive boycotts (they have a list of things to boycott as long as your arm), telling people how they should think and feel and what they should wear and eat. Around lunchtime they sneak into grade schools and whisper to the poor impressionable innocent youngsters there, *You know that sandwich that Mommy packed for you? Well, I know you love your mommy very much, but do you know that substance in your sandwich once had a mommy and a life too, and it wanted to live that life just as much as you want to live yours*. . . . (Animal rights activists only pretend to like children—they really don't think they're any more important than an earthworm or a piglet. What they really want to do is to give our children nightmares.) In the afternoon they indulge in more protests against circuses, zoos, and aquariums, which offer innocent family fun and/or education. They'll also attack the medical establishment whenever they can. (They dislike the sick because it is the sick who will benefit most from the data wrung from the research on intact live animal subjects, subjects that of course, if they're going to be of any use at all, do not stay intact for long. If animal rights activists had any real guts, they'd protest against who they're really against— tiny, sick babies, people with cancer and crippling diseases.)

As the day wanes, they go home and work for a while on the "Transitions" section of their underground zine, writing loopy obits like *Emily the hen enjoyed two years with her human friend Sally before taking spirit form and leaving this material plane*. . . . Evening finally comes and they sit eating their tofu burgers in the messy house they share with six dogs and eleven cats, watching inflammatory videos showing leghold traps going off and nailing the damnedest things; showing mutilated lab animals and terrified stockyard animals prior to stunning and skinning—they sit watching, watching simian horror, avian horror, equine horror, pound horror, trash cans full of euthanized dogs and cats in humane shelters, horses tied to trees and shot for bear bait—stuff a normal person would never want to look at—and then the last straw, the unfortunate footage of that elephant in Honolulu, Tyke, who escaped from the circus and was shot over and over again by police on the street, still had its little hat and bangles on and everything.... *Crazed* by such blatant propaganda, they rush out into the night, oblivious to the fact that they have run over a woman soliciting money—nickels, dimes, anything is welcome—for her ill, only son who needs a pig's heart valve if he's going to make it to his sixteenth birthday. Without pity or mercy or common decency, the activists in an orgy of vandalism smash the windows of fur stores and glue the door locks of businesses where hard-working taxpayers are trying to sell Heavenly Hams or leather sofas. Worse (they're quite deranged by now and should be captured and incarcerated) they break into a lab (a federal offense) where researchers have worked for years and years carefully, scientifically, decapitating cats, decerebrating dogs, and burning rabbits so that you and your children and your children's children can enjoy a better life....

Behold, the monster! The Animal Rights Activist. A mean-spirited, misinformed NUT, antitechnology, anti-science, antihuman with a bizarre agenda of rights for animals (and what the hell does that mean exactly, what kind of rights? the right to vote? the right to a good education? the right of a doggy to its own water dish? How about the right not to be nutted at the vet's? The right to die? (A right by the way that's long been accorded to them.) The animal rights activist is very, very misguided if he thinks he can dismantle a sensible, progressive, cutting-edge society that offers its citizens 40,000 different formulations of pesticides and 205,000 different types of prescription drugs alone, a humane society that is already committed to minimizing any purported suffering experienced by the 2.7 million animals trapped each year for their fur, the 7 billion animals slaughtered each year for food and the 20 million excluding rodents sacrificed for the purposes of science.

Industry; the factory farm; chemical, biotech, and pharmaceutical companies; et cetera—all of which have only the interests of everybody at heart—would also like to point out exactly who the hell these animal people are. They're leftists. They're totalitarians. They're panty-waisted pantheists. They're fanatics. They're hopelessly middle-class individuals with too much time on their hands. They're dangerous radicals. The comparison they frequently make between the human slavery of yesteryear and the treatment of animals today is incredibly offensive to blacks. The Holocaust imagery they toss around is incredibly offensive to Jews. Feminists should be deeply suspicious of their demand to give animals the right to have a life. The Church should be outraged at the suggestion that like us animals possess souls. Veterinarians should disavow them, and children should be protected from their stories. (We encourage the dissection of frogs in the classroom if the teacher can find such amphibians anymore....) Vegetarians should distance themselves from them completely. Vegetarianism doesn't have to be a wacky pseudo-ethical choice; it can, it should, be just a harmless personal preference.

The establishment that preserves and protects our lifestyle would also like to express the wish that the animal rights movement would just go away. Or at least become as docile, bland, and ineffective as the mainstream environmental movement, a movement that has been effectively neutralized in less than thirty years.

...

Animal rights groups are out in the big utopian lonely thinking paradisical. They have never been embraced by the increasingly corporate environmental community. Greenpeace, a once tough and charismatic organization, has been caught in exaggeration and lies and is now so muddy-minded that it supports legislation that will bring back tuna netting practices that proved so fatal to dolphins. The Nature Conservancy swaps land and triages habitat with unseemly tax-write-off vigor. Defenders of Wildlife, raking in the dough from wolf lovers in their highly publicized reintroduction program (and beginning to overdo it with those pictures of brave caring men in their mackinaws crouched over anesthetized wolves), is at the same time supporting increased government culling of the animals. Defenders also joined the World Wildlife Fund, the Environmental Defense Fund, and the National Wildlife Federation, among others, in their support of NAFTA, which begot GATT, which created the World Trade Organization, an extremely earth-unfriendly juggernaut. The Audubon Society, also a cheerleader for GATT, is the most reactionary of them all, but what could one expect from a group named after the premier avian slaughterer of his time. . . . ECOWIMPS all, as their duped and disappointed supporters are discovering. Yet even the far from ecowimpy Earth First! has never entangled itself in the briar patch that is animal rights. Farm animals to them are the problem. *Shoot cows not bears,* EF! exhorts in its typical Dada way. The Wild is everything, the Tame holds very little interest.

Jim Mason in his book *An Unnatural Order* makes the point that the conservation and environmental movement has always avoided the Animal Question. (You might as well call it the Animal Problem.) Though they call for radical changes in our worldview regarding Nature, the scholars and philosophers of environmentalism, the Deep Ecologists, somehow never mention the Animals, instead preferring the remoteness of discussions about trees or the abstractions of biodiversity and species. The call for a new, less anthropocentric ethic, an awakening, never acknowledges the reality, the *difficulty* of the animals in a new order. Changing the status of animals is discounted as a peripheral, even unworthy, concern.

But if environmentalists (when they're not out compromising) spend too much time contemplating their Gaia navels, society, in general, seems willing to consider the down-to-earth plight of the animals. That is, people seem to want to be kinder to animals even as they continue to use them and eat them and relocate them when it's time to build a vacation home. People support the animal rights movement to the degree they believe it is concerned with animal welfare. And their compassion and concern can be counted on to a point. But the perception about activists is that they go too far. Normal people are fond of animals and disapprove of wanton cruelty but keep their priorities in order. When a hurricane drowned 2 million

assembly-line-produced turkeys, chickens, and hogs in North
Carolina, the graphically revealed gothic methods of modern
animal husbandry was not the news, it was emphasized, it was
the possible contamination of the public water supply from
overflowing wastepits. (Hog farmers, you realize, have to raise
more hogs faster because there's less demand for pork.)

Normal people put people first. They poison rats, like fried
chicken, buy their dogs cow hooves as treats, and keep the
birdbath filled. When a dog was found bound and gagged with
electrical cord and set on fire in Miami, people contributed
money to a reward fund for the apprehension of his killer. A
few people contributing a little money would have been *normal*, but hundreds of people contributed a considerable amount
of money, which made them peculiar. *The Miami Herald* was
puzzled: "It exceeds the $11,000 offered by law enforcement
agencies for the capture of a serial killer who beats and burns
homeless women here."

When a seventeen-year-old with cancer wanted to go to
Alaska and kill a Kodiak bear and was sent to do just that,
thanks to the generosity of the Make-a-Wish Foundation, it
set off what the papers referred to as "an animal rights furor."
The extent of that furor caused others to be more "objective"
about the situation, saying things like *Hey, it'll make the poor
kid happy, it's a legitimate wish, and it's something he can do with
his dad.*

When boys on a high school baseball team in Texas killed
a cat by battering it with their bats, stuffing it in a bag and
running over it with a pickup truck because it had taken to
hanging around and soiling the pitcher's mound, the "animal

people" were outraged and demanded that the boys be kicked off the team. Such disapproval "bewildered" the youths. "It was just a stray cat," one of the coaches said. "We all did things to cats when we were young. Some people think a cat is more important than a boy."

Almost everyone has heard the remark made by People for the Ethical Treatment of Animals (PETA)—"A rat is a pig is a dog is a boy"—and it has been used with considerable success to discredit the animal rights movement. PETA's actual statement was "When it comes to having a nervous system and the ability to feel pain, hunger, and thirst, a rat is a pig is a dog is a boy." Even addressing the statement as intended has resulted in a not-so-edifying debate about suffering. Do animals suffer or don't they? And if they do (they certainly seem to), does that *ability*, rather than speech or reasoning, give them rights? (One of the more remarkable philosophic arguments against granting animal rights is that they have no sense of morality—we can't act morally toward them because they can't act morally back.) Suffering aside, when people care *too much* about animals, it's suspected that somewhere, somehow, some person is being deprived of generic love and support and attention because of it.

A high-school baseball coach is probably not up to the debating dazzle, say of Dostoyevsky's Grand Inquisitor, but it is remarkable how often this argument, puffed up a bit, is used to defend the uses of animals in research. Laboratory animals, although not deemed *un-animals*, have been transformed semantically into animal "models." Like "food" animals, they qualify for very little protection under the Animal Welfare

Act. Blinding has long been a popular procedure in the lab, as are any and all "deprivology" studies. Of endless interest is the study of an animal's reaction to unrelieved, inescapable pain. The procedures, of course, are never cruelty but science—they may result in data that might be of some use to us sometime. So baboon heads are bashed in to (ostensibly) help the thousands of people who suffer head injuries each year; calves are given heart attacks to (possibly) be of use to people who are going to have heart attacks; dogs are deliberately poisoned with insecticides to help children who are accidentally poisoned with insecticides (a rat is a dog is a boy when it suits science's purposes); other dogs are tormented into states of trauma, into states of "learned helplessness" into "psychological death," to give insights into human depression (maybe) or just to provide grist for a thesis. Other types of experiments serve different purposes. There are the voodoo and leech variety—"Cats Shot In The Head Usually Die, Tulane Study Finds." There are the experiments that merely satisfy scientific "curiosity." There are the *let's do this and see if something interesting happens* kind and there are the *wow this stuff vaporized this puppy's skin right down to the bone—I wonder if it will take the rust off lawn furniture with no mess* kind. Other experiments serve merely to confirm prior conclusions—to verify previously known LD (lethal dose) levels, for example. LD tests, used by industry to determine the toxicity of floor waxes and detergents (pumped directly into the animals' stomachs through tubes) end when half of the participants in a test group die. Animals almost never leave laboratories alive. They keep going into more corrosive tests or endure more invasive procedures until they succumb or until, their bodies unable to provide even the most senseless data, they're "humanely destroyed."

Of all lab animals, the chimpanzee is the most popular. The chimps—humankind's closest relative—are infected and maimed and killed for us, for the possible advantage to us, because they're so much like us. They possess 98.6 percent of the same DNA, the same genetic material. That missing 1.4 percent allows them to be vivisected on our behalf. If it weren't for that lucky-for-us 1.4 percent, they wouldn't be able to be used as experimental subjects because they'd be just like us, and medical advancement would be completely thwarted. It would come to an absolute standstill, it would, in the words of a doctor writing in the *New England Journal of Medicine*, "spell complete stymie."

So in our country's finest universities (as well as in some of our just so-so ones) researchers, not to be stymied, are still making test animals "hot" with deadly diseases and screwing bolts in their heads. They're still performing cataract surgery on healthy ones, then giving them different rehabilitation treatments, then killing them and dissecting their brains to see which treatment produced the best result within the visual cortex. (Monkeys can't be trained to read eye charts.) But perhaps most dramatically of all, researchers throughout the '80s and '90s tried to give chimps AIDS. Anyone opposed to animal research was damned as being against finding a cure (only a matter of time) for one of the most dreadful scourges of modern times. Chimps became the Poster Primates in the fight against AIDS, employed as a symbol against the animal rights movement.

But after infecting more than one hundred chimps with HIV, scientists became increasingly frustrated with their lack of success. They could kill a monkey by destroying his own simian immune system, but they simply could not give him AIDS. Vaccines, created by injecting weakened simian viruses into chimps, also proved to be disappointing, merely creating the fatal animal disease they were intended to prevent. Now, after twenty years, scientists no longer consider our fellow primates to be "good" models for AIDS research. The intensive breeding of chimps for research, blessed by the National Institutes of Health, has resulted in eighteen hundred "excess" chimps. There were always researchers eager to infect them, but few—actually none—are interested in caring for the animals for their natural life span, about forty-five years.

Artificially induced diseases in animals practically never result in a cure for those diseases that can be applicable to humans, although the risk of strange new species-jumping sicknesses grows stronger every day. Misleading monkey experiments delayed an effective polio vaccine for decades. Successes in human kidney transplants, blood transfusions, and heart bypass surgery all resulted when doctors ignored the baleful results of experiments on dogs and used human material. Guinea pigs die when injected with penicillin. Thalidomide was found to be safe, safe, safe for rodents; so was Opren, an arthritis drug that caused fatal liver toxicity in a number of human patients before it was taken off the market. Animal tests for new drugs do not predict side effects in humans up to 52 percent of the time. The National Cancer Institute stopped testing anticancer drugs on animals in the mid-1980s, finding toxicity results useless and misleading. (For instance, 46 percent of substances deemed carcinogenic in mice are noncarcinogenic in rats.) It is common now upon the announcement of any implication wrung from animal research for the researchers to publicly caution against using the findings to make conclusions about human disease or behavior.

And then there is the smoking debacle. The tobacco industry was able to deny a link between cigarette smoking and lung cancer for decades because many thousands of dogs, monkeys, rabbits, and rats, fitted with masks and placed in "smoking chambers" or immobilized in stereotaxic chairs with tubes blowing smoke down their windpipes could not be encouraged to develop carcinomas. In a more perfect world these animals all would have gotten lung cancer, thereby savings millions of human lives. They would have given a future with hope to an entire generation . . . (but those stubborn creatures didn't do it).

Such a dismal record of nonachievement explains why enthusiasts of animal experimentation have to devote a portion of their energies and profits to public relations (Vivisectors Do It For *You!*). One of their most effective strategies is to enlist people in the war not against disease but against animal rights supporters. . . . *I just want to thank the National Institutes of Health because I wouldn't be in remission today had it not been for all those dead dogs,* some tearful housewife says to a congressional committee; or . . . *I guess those animal people would like to make me feel ashamed to say this, but I think the life of my baby is more important than that of a research kitten. . . .* Parents' terror of the mysterious Sudden Infant Death Syndrome was manipulated shamelessly with the *cure-is-dependent-upon-animal-research* mantra until the precipitous recent drop in infant deaths was attributed to the simple act of putting babies to bed on their backs instead of on their stomachs, a change in custom that has been described as "one of the simplest and most effective public health interventions ever." (An ounce of prevention may be worth a pound of cure, but it's certainly not something drug companies are interested in.)

Yet when caught in the manipulation or misrepresentation of data or the exploitation of fears and hopes, science and medicine, hand in hand, retreat to the high ground, professing to be saddened by a public that so "fears expertise," that is so unobjective and so ignorant, that is so shockingly unaware of the nature of science (being wrong is a constant feature of scientific method), of what it can do successfully (potentially everything) and what it cannot do (resolve society's ethical qualms, or make value judgments).

Between competing notions a nonrational incommensurability lies, thus a switch from one belief or manner of action to another can never be achieved by logic alone. Major changes in thinking are frequently not incremental but occur all at once. In the book *Scientific Revolutions*, author Thomas Kuhn likens such change to the gestalt shift: "These flashes of intuition resemble instantaneous electron orbital changes. They are never in transition. They are here. Then they are . . . there."

Animal experimentation seems to be bloodily quaint baggage to carry into the twenty-first century and vivisection a most primitive method of discovery. Still, it does seem that something like instantaneous electron orbital exchange in the mass mind is required to make the vivisector's work totally unacceptable to society. The Animal Liberation Front breaks into labs, damages equipment, and frees animals, all to great notoriety and accusations of terrorism, but its raids often provide irrefutable proof of researchers' barbarism. The ALF stole films from the University of Pennsylvania's head injury lab that showed baboons in vises getting their heads smashed while researchers chortled. The National Institutes of Health had called the Pennsylvania lab "one of the best in the world," but the federal government cut off funding after the improperly acquired film was made public. And if the young, rosy-cheeked commandos from the ALF with their bolt cutters and black ski masks broke into all the labs, emptied all the cages, and carried the deliberately diseased and wounded and moribund animals in litters down Constitution Avenue next New Year's Day, and if the event was featured on all the television networks in vivid and unrelenting repellent close-up, horrifying a majority of the viewing public, who would find the sight virtually pornographic, what would happen in this world of laws?

Well, all the rosy-cheeked, idealistic, No Compromise! ALF members would be jailed without bond pending trial on criminal trespass and theft charges, and the animals, the stolen property, would be returned to the labs for disposal. Prior to the return, the labs would have to rely even more than usual upon B dealers who are licensed by the Department of Agriculture to provide pound and other "random source" animals for them, "random source" frequently being lost or stolen pets. (Medical researchers prefer former pets over other animal sources since they are easier to control, more trusting and obedient. . . .) Congress, urged on by a distraught citizenry, would debate what, or whether, changes would be necessary in the wording of the Animal Welfare Act (which surely must have the welfare of animals as its intent—otherwise, why is it called that . . .) because, as it is, the wording does not prohibit any type of experiment or procedure that can be performed on animals in laboratories and makes clear that the government cannot interfere with the conduct or design of those procedures. There would be much talk about wording, and strengthening the existing intent of the wording. The language of rights—practically the only ethical language we speak in this country—would not be spoken. Instead, humane treatment would be interpreted and described, as would suffering, and pain. Researchers and other interested parties would argue that animal suffering is an emotionally charged term that can't be defined as a reality, and that any attempt to define it would be biased on the side of sentimentality or sympathy and would be intellectually unverifiable. The "animal people" would speak about intentional cruelty inflicted upon sentient beings, and the researchers would say: *It's not cruelty, it's science. I suppose you'd prefer us to experiment on severely retarded people instead—well, we're not going to do that, that would be morally reprehensible and, of course, against the law. We have laws in this country, you know. . . .*

And so it would go. Maybe bigger cages for the beasts would be required. Maybe daily water would be legislated (except in those cases where the denial of water was the point of the experiment). But the wrong would not be addressed because no right would have been established.

• • •

There are thousands of animal-advocacy organizations in the United States, with millions of members. Feral cats, wild horses, greyhounds, fowl, bats, as well as the more dramatic gorillas, pandas, and dolphins, all have their devoted protectors, and various methods are used to win public sympathy for them. But many advocates are working for more humane treatment of animals and would prefer not to argue the rights issue at all. To argue that a monkey has the right not to have his arms cut off in an experiment is far different than arguing that a thirsty horse should be given water before his journey to becoming dog food. It is one thing to show up as a carrot at the county fair toting a placard that reads, "Eat Your Veggies, Not Your Friends," and quite another to find a convincing language with a commanding legal basis that liberates animals from "thinghood." It's one thing to rescue individual animals from the slaughterhouse by buying them, as the group Farm Sanctuary does (you can sponsor a sheep for $20 a month, a duck for $8), and quite another to argue that raising animals for slaughter is morally unacceptable.

It's easier to have a little brown rat as a pet (very affectionate) or even to make cruelty-free stock investments than it is to wade into real rights talk and tempt flake status. Rights is radical and abolitionist, welfare is conservative (the word to some extent has already been co-opted and absorbed by the status quo) and reformist.

The Humane Farming Association may be radical in its methods—in one case, members of the group slipped into a slaughterhouse and stole calves' eyeballs to test for the toxic drug clenbuterol—but its purpose is to make farming more environmentally responsible and to protect and enlighten the consumer. The Humane Society of the United States has become politically sophisticated at lobbying and promoting ballot initiatives (patiently, patiently, which causes more obstreperous groups like the Animal Liberation Front to behave impolitely—like driving up to a McDonald's with a dead cow in the back of a pickup truck and a sign saying "Here's Your Lunch"). Even so, HSUS, traditionally considered kitty and doggy moderate, has, since 1980, claimed that there is no basis for maintaining a moral distinction between the treatment of humans and other animals, a view quite extremist in its implications.

Welfare groups have been laboring on behalf of the animals for some time—the American Society for the Prevention of Cruelty to Animals and the American Anti-Vivisection Society are both over a hundred years old—but the rights movement took off only in 1973, when the *New York Review of Books* published an unsolicited review of the book *Animals, Men, and Morals* by Roslind and Stanley Godlovitch and John Harris. The writer was the Australian philosopher Peter Singer, who expanded his article into the ideological classic *Animal Liberation*. PETA, founded by Ingrid Newkirk and Alex Pacheco in 1980, is the group that perhaps best personifies the rights movement, for it broke tactical ground in 1981 with a daring legal action that attempted to prosecute a researcher for animal cruelty. Pacheco volunteered as an assistant to a Dr. Edward Taub at the Institute of Behavioral Research in Silver Spring, Maryland, with the intention of secretly documenting conditions in an "ordinary" lab. Taub had been surgically crippling primates to monitor the rehabilitation of impaired limbs for many years, apparently suspending his efforts only long enough to write proposals for federal grants that would, and did, allow him to continue his labors. Pacheco and PETA got a precedent-setting search warrant from a circuit judge, and police raided the filthy lab and confiscated seventeen monkeys, as well as Taub's files and a monkey's severed hand that the less-than-charismatic researcher kept on his desk as a paperweight. Although the rights of the mutilated primates could not be argued, as those rights had never been established, Taub was found guilty of cruelty to animals by a jury. The conviction was overturned on appeal when the court ruled that state statutes did not apply to research conducted under a federal program. Since then, additional daring cases have been won, only to be lost on appeal, and the cases that are won involve animal cruelty or welfare, never the rights of an animal, for of course an animal has no rights; an animal has no standing in a court of law. The injuries to a person's "aesthetic interests" can be judicially recognized (I am offended by seeing spotted owls mounted on the hoods of logging trucks . . .), but an animal's interest in continuing to exist cannot.

• • •

The animal people need their day in court on the rights issue, and groups such as the Animal Legal Defense Fund are seeking to find, try, and win the perfect case—the case that will take animals out of the realm of property and grant them legal status of their own. The plaintiff will undoubtedly be a chimp. The chimpanzees' ability to be trained in sign language, and their further ability to use that language to express their fears and needs, could provide the scientific basis for the argument that they deserve the same freedom from enslavement that most humans now enjoy. Peter Singer's latest philosophical effort is the Great Ape Project, a rhetorical demand for the extension of the "community of equals" to include all the great apes: human beings and "our disquieting doubles"—chimpanzees, gorillas, and orangutans. The rights of life and freedom from torture and imprisonment would be granted to these animals, and then, possibly, would trickle down to those our less disquieting doubles.

Sometimes a number of animal people gather together as they did recently for a "World Congress" at the cavernous USAir Arena in Landover, Maryland, just outside Washington, D.C. The arena can hold eighteen thousand people, and it was not full. The dilemma of the animals still lacks the drawing power of a mediocre rock band. Of course, animals can never be called upon to do a star turn on the movement's behalf—that would be anathematic to the whole precept. So only people were there, about three thousand of them. There were three days of speeches. The crowd was attentive. The speakers were calmly impassioned, well spoken, nicely dressed, well prepared; they politely restricted themselves to the time allotted. Nobody screamed, *We've got to stop dressing up as carrots!* or *Whose idea was it to petition the town of Fishkill to change its name? It made us look like morons!* The importance of unity was stressed. But there are so many methods, so many concerns, so many didactics . . . so many mundane and divine and extravagant ways to work for the animals. Managing "colonies" of feral cats by setting up feeding stations (Alley Cat Allies) seems less noble than trying to save a silverback gorilla (The Biosynergy Institute) from the stewpot, but it would be impolite to say so at the big meeting of the year. The group the Nature of Wellness—with its witty baby-in-a-bonnet newspaper ads ("Most People See A Beautiful Healthy Child. We See A Cure For Feline Leukemia"), which ridicule the premises and promises of animal testing—probably runs a more intellectually engaging campaign than the person who decides to show up outside Macy's some summer afternoon pretending to be a fur-bearing animal with its leg in a trap, but no one's going to question the latter's devotion. A fifth-grade teacher from Charleston who has her class read *Black Beauty* and then write an essay "If I Were an Abused Carriage Horse, What Would My Three Wishes Be?" leads a different, and probably less accident-prone life than the individual from Colorado who goes out to harangue the hunters on opening day. The feminists with their earnest semantic quibblings and Internet horror stories (alt.sex. bestiality . . . *One person described having sex with stray dogs and then dropping them off at animal shelters . . .*), equating everything with the subjugation of women, may not be quite as helpful in the long run as the lawyers strategizing in the Animal Legal Defense Fund, but no one's going to tell them to lighten up. The members of the Animal Rights Foundation of Florida who brought about the cancellation of a liposuction surgery workshop using live pigs can be just as proud of their work on behalf of the animals as the animal law lawyers, theorists, and essayists Steven Wise and Gary Francione, who teach at Harvard and Rutgers, respectively. Everyone here is working, working, working for the animals in the hope that everything, somehow, is going to add up and that the animals will be . . . saved.

There is no limit to the things an animal rights activist can worry about in this world. And it became clear as the groups were heard and topics covered and enemies identified that whenever a battle is won, a victory claimed for the animals, it doesn't stay a victory for long. It's either nondefinitive, or it's superceded by something worse.

Two great successes for the movement involved the fur and the cosmetic industries. The wearing of fur was discredited through the tactics of howling insult. *Corpse Coat!!* activists would scream at the slightest opportunity, or they would solicitously ask of some fur wearer, *How did you get the blood off that?* Then they'd go out and paint "SHAME" and "DEATH" all over furriers' windows. Education followed by organized consumer boycotts encouraged cosmetic companies to pretty much eliminate animal testing. (*Mommy, is it true that they blinded hundreds of white bunnies to make this pretty soap?*)

But the fur industry is still around, hoping for government subsidies to boost export sales and counting on a new wave of designers—there's always a new wave—who believe the trend gurus' predictions of a "fur renaissance fueled by a growing interest in luxury investments" and are churning out the beaver capes, the burgundy ponyskin jackets, and the acid green sable barn jackets. And some of the big names in the beauty industry—Helene Curtis, Chesebrough Pond's— continue to test on animals. Overall, the use of animals in research could very well be increasing—who knows? Corporate monoliths such as Procter & Gamble and Bausch

& Lomb never stopped animal testing; the Department of Defense could still be cutting the vocal cords of beagles and testing nerve gas on them. The DOD doesn't have to release any figures at all, and research facilities in general enjoy institutionalized secrecy and seldom have to provide real numbers to the public. The Silver Spring monkeys that PETA pried from researcher Taub almost two decades ago became wards of the National Institutes of Health and were only recently put to death (humanely, of course) after a final set of experiments in which the tops of their skulls were removed and their brains repeatedly pierced with electrodes. Augustus, Big Boy, and Dominion had already lost the use of their limbs in previous experiments when government-financed scientists at Tulane had severed the nerves to their arms and shoulders.

No, there's little cause for real happiness among the animal people and scant opportunity for self-congratulation. (Peter Singer says about *Animal Liberation*, "When I wrote it, I really thought the book would change the world. All you have to do is walk around the corner to McDonald's to see how successful I've been.")

Public awareness and revulsion are often raised only to fade or be circumvented. Commercial whaling has never been outlawed; the clubbing of seals has resumed; trade in exotic species is brisk; bills pending in the House and Senate would allow tuna fishermen to again use netting methods that would kill thousands of dolphins and still be able to label their product as "dolphin safe" (sort of a joke on the little kids). The USAir Arena itself, so vast and impersonal, so disconcertingly inert, only emphasized the gargantuan task the animal people had taken on, and the gaunt specter of hopeless helplessness appeared more than once. Between speeches, people would wander out to the encircling satellite area and line up for the beyond-veggie vegan food that the arena's concessionaires were serving up with a certain amount of puzzlement. The Franks-a-Lot stand was sensibly shuttered. (The animal people are vegetarians. They'd better be if they don't want to be accused of being hypocritical. Of course, by being unhypocritical, they can be accused of being self-righteous.)

On the fourth day of the World Congress there was a March for Animals, from the Ellipse up Constitution Avenue to the Capitol. Any parade watcher who had expected to see animals (as though animal rights activists were all Episcopalians going to some fun-filled blessing of the beasts at church and taking their rats and snakes and dogs and burros and pigs with them) would have been disappointed. And anyone who had expected to see eccentricity incarnate in the marching crowd would have been disappointed as well. They appeared to be ordinary, caring, middle-class Americans marching for justice. Yet has any group in this country ever had such an extremist agenda, based utterly on non-self-fulfillment and non-self-interest? The animal people are calling for a moral attitude toward a great and mysterious and mute nation. Their quest is quixotic; their reasoning, assailable; their intentions, almost inarticulateable. The implementation of their vision would seem madness. But the future world is not this one. Our treatment of animals and our attitude toward them are crucial not only to any pretensions we have to ethical behavior but to humankind's intellectual and moral evolution. Which is how the human animal is meant to evolve, isn't it?

Electric Chair

THE PRACTICALITY OF DEATH BY ELECTROCUTION WAS PROVEN in Thomas Edison's odd Menlo Park Laboratory in 1888. Edison allowed an engineer named Harold Brown to fiddle around in the dynamo room, pushing cats and dogs onto a metal plate electrified with alternating current. Edison did not care for alternating current. He didn't understand it. He liked direct current. But a competitor, George Westinghouse, was proving to be successful combining high-voltage AC with a transformer, transmitting farther and cheaper than Edison could with his DC methods. Westinghouse had had some unfortunate accidents, however, and Edison never tired of adding up the competition's volts ("HOLY MOSES!" he wrote) and emphasizing their dangers in letters and memorandums.

In order to demonstrate how awful and potentially deadly AC was, Edison allowed Brown to experiment, and after sizzling some fifty animals, Brown declared AC a perfect electrical medium for efficient extermination and inaugurated a campaign against Westinghouse as a "merchant of death." Brown also devised an "electrical cap and shoes" and sold them, along with three AC dynamos, to the state of New York for $8,000. The state, which had been seeking a more modern alternative to hanging, was enthralled. State officials added a chair and some straps and wanted to call it something like *electromort* or *dynamort*. Edison slyly suggested that they call it Westinghousing and continued to lead more animals to their death—dogs, calves, a horse, and an elephant—to prove AC's dangers on one hand, and awesome effectiveness on the other. "I have not failed to seek practical demonstration," Edison wrote. "I have taken life—not human life—in the belief that the end justified the means."

Hawk

GLENN GOULD BATHED HIS HANDS IN WAX AND THEN THEY felt new. He didn't like to eat in public. He was personally gracious. He was knowledgeable about drugs. He loved animals. In his will, he directed that half his money be given to the Toronto Humane Society. He hated daylight and bright colors. His piano chair was fourteen inches high. His music was used to score *Slaughterhouse-Five*, a book he did not like. After he suffered his fatal stroke, his father waited a day to turn off the respirator because he didn't want him to die on his stepmother's birthday. When Glenn Gould wrote checks, he signed them Glen Gould because he was afraid that by writing the second n he would make too many squiggles. He took prodigious amounts of Valium and used makeup. He was once arrested in Sarasota, Florida, for sitting on a park bench in an overcoat, gloves, and muffler, attire the police deemed suspiciously unsuitable for the climate. He was a prodigy, a genius. He had dirty hair. He had boring dreams. He probably believed in God.

My mind said, *You read about Glenn Gould and listen to Glenn Gould constantly, but you don't know anything about music. If he were alive, you wouldn't have anything you could say to him. . . .*

A composer acquaintance of mine dismissed Glenn as a performer.

Glenn Gould loved the idea of the Arctic, but he had a great fear of the cold. He was a virtuoso. To be a virtuoso you must have an absolutely fearless attitude toward everything, but Glenn was, in fact, worried, frightened, and phobic. The dogs of his youth were named Nick and Banquo. As a baby, he never cried but hummed. He thought that the key of F minor expressed his personality.

You have no idea what that means, my mind said. *You don't really know what it is he's doing. You don't know why he's brilliant.*

He could instantly play any piece of music from memory. On the whole he did not like works that progressed to a climax and then to a reconciliation. The Goldberg Variations, which Glenn is most widely known for, were written by Bach for harpsichord. The story goes that Bach was visiting one of his students, Johann Goldberg, who was employed by a Count von Keyserling, the Russian ambassador to the court of Saxony. The count had insomnia and wanted some music that would help him through the dark hours. The first notes of the Goldberg Variations are inscribed on Glenn's tombstone.

My dog rose from his bed and walked beneath the table, which he barely cleared. He put his chin on my knee. He stood there for a few moments, not moving. I could see nothing but his nose. I loved kissing his nose. It was my hobby. He was a big black German shepherd with accents of silver and brown. He had a beautiful face. He looked soulful and dear and alert. He was born on October 17, 1988, and had been with us since Christmas Day of that year. He was now almost nine years old.

He weighed one hundred pounds. His name was Hawk. He seemed to fear nothing. He was always looking at me, waiting for me. He just wanted to go where I was going. He could be amusing, he had a sense of humor, but mostly he seemed stoic and watchful and patient. If I was in a room, he was in that room, no other. Of course we took long walks together and many cross-country trips. He was adept at ferry crossings and checking into motels. When he could not accompany me, I would put him in a kennel, once for as long as two weeks. I felt that it was good for him to endure the kennel occasionally. Life was not all good, I told him. Though mostly life was good. He had a series of collars over the years. His most recent one was lavender. He had tags with his various addresses and phone numbers on them and a St. Francis medal with the words "protect us." He had a collection of toys. A softball, and squeaky toys in the shapes of a burglar, a cat, a shark, a snowman, and a hedgehog that once made a snuffling noise like a hedgehog but not for long. They were collected in a picnic basket on the floor, and when he was happy, he would root through the basket and select one. He preferred the snowman. His least favorite was a large green and red toy—its shape was similar to a large bone, but it was an abstraction, it lacked charm. Hawk was in a hundred photographs. He was my sweetie pie, my honey, my handsome boy, my love. On the following day he would attack me as though he wanted to kill me.

Regarding life, it is much the best to think that the experiences we have are necessary for us. It is by means of experience that we develop and not through our imagination. Imagination is nothing. Explanation is nothing. One can only experience and somehow describe—with, in Camus's phrase, *lucid indif-ference.* At the same time, experience is fundamentally illusory. When one is experiencing emotional pain or grief, one feels that everything that happens in life is unreal. And this is a right understanding of life.

I loved Hawk and Hawk loved me. It was the usual arrangement. Just a few days before, I had said to him, This is the life, isn't it, honey? We were picnicking on Nantucket. We were on the beach with a little fire. There was a beautiful sun-set. Friends had given my husband and me their house on the island, an old farmhouse off the Polpis Road. Somehow, on the first night at the house, Hawk had been left outside. When he was on the wrong side of a door, he would never whine or claw at it; he would stare at it fixedly. I had fallen into a heavy sleep.

I was exhausted. I was always exhausted, but I didn't go to a doctor. I had no doctor, no insurance. If I was going to be very sick, I would just die, I thought. Hawk would mourn me. Dogs are the best mourners in the world, as everyone knows. In my sleep, in the strange bed in the old farmhouse, I saw a figure at the door. It was waiting there clothed in a black garbage bag and bandages. I woke and without hesitation got up and went to the door and opened it, and Hawk came in. Oh I'm so sorry, I said to him. He settled down at the foot of the bed with a great comfortable sigh. His coat was cool from the night. I felt that he had tried to project himself through to me, that he had been separated from me through some error, some mis-understanding, and this, clearly, was something neither of us wanted. It had been a bad transmission, but it had done the job and done it without frightening me. What a resourceful boy! I said to him. Oh there are ghosts in that house, our friends said later. Someone else said, You know, ghosts frequently appear in bandages.

Before Hawk, I had had a number of dogs who died before their time, from grim accident or misfortune, taken from me unprepared in the twinkling of an eye. *Shadrach, Nichodemus, Angel* . . . Nichodemus wasn't even old enough to have learned to lift his leg. They were all good dogs, faithful. They were innocents. Hawk was the only one I didn't name from the Bible. I named him from Nature, wild Nature. My parents always had dogs too, German shepherds, and my mother would always say, You have to talk to a dog, Joy, you've got to talk to them. It ended badly for my mother and father's dogs over the years and then for my mother and father. My father was a Congregational minister. I am a Christian. Kierkegaard said that for the Christian, the closer you keep to God and the more involved you get with him, the worse for you. It's as though God was saying . . . *You might as well go to the fair and have a good time with the rest. Don't get involved with me—it will only bring you misery. After all, I abandoned my own child. I allowed him to be killed.* Christianity, Kierkegaard said, is related only to the consciousness of sin.

We were in Nantucket during *dies caniculares*, the dog days of summer, but it was a splendid time. Yet there was something wrong with me. My body had turned against me and was full of browsing, shifting pain. The pain went anywhere it wanted to. My head ached, my arms and legs and eyes, my ribs hurt when I took a deep breath. Still, I walked with Hawk, we kept to our habits. I didn't want to think about it, but my mind said,

*You have to, you have to do something, you can't just do nothing,
you know.* . . . Some days were worse than others. On those
days, I felt crippled. I was so tired. I couldn't think, couldn't
concentrate. Even so, I spent long hours reading and listening to music. Bach, Mahler, Strauss. Glenn thought that the
"Metamorphosen" of Strauss was the ultimate. I listened to
Thomas de Hartmann play the music of Gurdjieff. I listened
to Kathleen Ferrier sing Mahler and Bach and Handel and
Gluck. She sang the famous aria from Gluck's opera *Orfeo ed
Euridice*—"What Is Life." We listened to the music over and
over again.

Hawk had engaging habits. He had presence. He was
devoted to me. To everyone, this was apparent. But I really
knew nothing of his psychology. He was no Tulip or Keeper or
Bashan who had been analyzed by their writers. He knew sit,
stay, down, go to your place. He was intelligent, he had a good
memory. And surely, I believed, he had a soul.

• • •

The friends who had given us the house on Nantucket insisted
that I see a doctor about my malady. They made an appointment for me with their doctor in New York. We would leave
the Island, return to our own home in Connecticut for a few
days, then put Hawk into the kennel and drive into the city, a
little over two hours away.

I can't remember our last evening together.

On the morning my husband and I were to drive into the
city, I got up early and took Hawk for a long walk along accus-
tomed trails. I was wearing a white sleeveless linen blouse and
poplin pants. My head pounded, I could barely put one foot

ahead of the other. *How about lupus?* my mind said. *How about rheumatoid arthritis?* Well, we'll know more soon. . . . We drove, then, to the kennel. It's called Red Rock, and Hawk had been there before; they liked him there, he'd always been a gentleman there. When we drove in, Hawk looked disconsolate yet resigned. I left him in the car while I went into the office. I was looking for Fred, big, loud, gruffly pleasant Fred, but he didn't appear. One of his assistants did, a girl named Lynn. Lynn knew Hawk. He's only going to be here for one night, right? Lynn said. I went out to get him. I put the leash on him, his blue, rather grimy leash, and he jumped out of the car and we walked into the office. Lynn had opened another door that led to a row of cement runs. We stood in that doorway, Hawk and I. All right then, I said. I was bent forward slightly. He turned and looked at me and rose and fell upon me, seizing my breast. Immediately, as they say, there was blood everywhere. He tore at my breast, snarling, I think, I can't remember if he was snarling. I turned, calling his name, and he turned with me, my breast still in his jaws. He then shifted and seized my left hand, and after an instant or two, my right, which he ground down upon, shifting, getting a better grip, always getting a better grip with his jaws. I was trying to twist his collar with my bleeding left hand but I was trying not to move either. Hawk! I kept calling my darling's name, Hawk! Then he stopped chewing on my hand, and he looked at me coldly. Fred had been summoned by then and had a pole and a noose, the rig that's used for dangerous dogs, and I heard him say, He's stopped now. I fled to the car. My blouse was soaked with blood, it was dripping blood. I drove home sobbing. I've lost my dog, I've

lost my Hawk. My mind didn't say anything. It was all it could
do to stay with me as I sobbed and drove, my hands bleeding
on the wheel.

I thought he had bitten off my nipple. I thought that when
I took off my blouse and bra, the nipple would fall out like a
diseased hibiscus bud, like the eraser on a pencil. But he hadn't
bitten it off. My breast was bruised black, and there were two
deep punctures in it and a long raking scratch across it and that
was all. My left hand was bleeding hard from three wounds.
My right hand was mauled.

At home I stood in the shower, howling, making deep ugly
sounds. I had lost my dog. The Band-Aids we put over my
cuts had cartoon characters all over them. We didn't take our
medicine cabinet very seriously. For some reason I had papered
it with newspaper pictures of Bob Dole's hand clutching its
pen. I put clean clothes on, but the blood seeped around the
Band-Aids and stained them too. I put more Goofy and Min-
nie Band-Aids on and changed my clothes again. I wrapped
my hand in a dish towel. Hawk's water dish was still in the
kitchen, his toys were scattered around. I wanted to drive into
the city and keep my appointment with the doctor; he could
look at my hand. It seemed only logical. I just wanted to get in
the car and drive away from home. I wouldn't let my husband
drive. We talked about what had happened as being *unbeliev-
able*. We hadn't yet started talking about it as being a *tragedy*.
I'll never see him again, I've lost my dog, I said. Let's not talk
about that now, my husband said. As we approached the city,
I tried to compose myself for the doctor. Then I was standing
on the street outside his office, which was on East Eighty-fifth

Street, trying to compose myself. I looked disheveled, my clothes were stained, I was wearing high-top sneakers. Some people turned as they were walking by and made a point of staring at me.

He was a cheerful doctor. He put my hand in a pan of inky red sterilizing solution. He wanted to talk about my malady, the symptoms of my malady, but he was in fact thinking about the hand. He went out of the examining room for a while, and when he came back, he said, I've made an appointment for you to see an orthopedic surgeon. This doctor was on East Seventy-third Street. You really have to do something about this hand, the first doctor said.

The surgeon was of the type Thomas Mann was always writing about, a doctor out of *The Magic Mountain*, someone whom science had cooled and hardened. Still, he seemed to take a bit of pleasure in imagining the referring doctor's discomfort at my messy wounds. People are usually pretty well cleaned up by the time Gary sees them, he said. He took X-rays and looked at them and said, I will be back in a moment to talk with you about your hand. I sat on the examining table and swung my feet back and forth. One of my sneakers was blue and the other one green. It was a little carefree gesture I had adopted for myself some time ago. I felt foolish and dirty. I felt that I must appear to be not very bright. The doctor returned and asked when the dog had bitten me and frowned when I told him it had been six hours ago. He said, This is very serious—you must have surgery on this hand today. I can't do it here, it must be done under absolutely sterile conditions at the hospital. The bone could become infected, and bone infections are very difficult to clear up. I've reserved a bed in the hospital for you and arranged for another surgeon to perform the operation. I said, Oh, but . . . He said, The surgery must be done today. He repeated this, with beats between the words. He was stern and forbidding and, I thought, pessimistic. Good luck, he said.

The surgeon at Lennox Hill Hospital was a young, good-looking Chinese man. He spoke elegantly and had a wonderful smile. He said, The bone is fractured badly in several places, and the tendon is torn. Because it was caused by a dog's bite, the situation is actually life-threatening. Oh, surely . . . I began. No, he said, it's very serious, indeed, life-threatening, I assure you. He smiled.

I lay in a bed in the hospital for a few hours, and at one in the morning the hand was operated on and apparently it went well enough. Long pins held everything together. You will have some loss of function in your hand, but it won't be too bad, the doctor said, presenting his wonderful smile. I used to kiss Hawk's nose and put my hands in his jaws in play. People in the hospital wanted to talk about my dog biting me. That's unusual, isn't it, they said, or, That's strange, isn't it, or, I thought that breed was exceptionally loyal. One nurse asked me if I had been cruel to him.

My hand would not be the same. It would never be strong and it would never again stroke Hawk's black coat.

When I was home again, I washed Hawk's dishes and put them in the cupboard. I gathered up all his toys and put them away too. I busied myself thinking I would bury all his things. Meanwhile, he waited at the kennel for me to come and get him, like I always had. I was taking Vicodin for the pain and an antibiotic. In a week I would begin taking another antibiotic and an anti-inflammatory drug for my malady. I lay about, feeling the pain saunter and ping through me. My arms felt like flimsy sacks holding loose sticks. If the sticks touched one another, there would be pain. I went back to listening to Glenn Gould and reading about Glenn Gould, which is what I had been doing when Hawk and I were last together. I played Glenn Gould over and over. Glenn never wanted to think about what his hands and fingers were doing, but as he grew older he became obsessed with analyzing their movements. He felt that if he performed with a blank face, he would lose his control of the piano. Frowning and grimacing gave him better control of his hands. My mind said, *You would not be able to defend or explain Glenn Gould to anyone who didn't care for him.*

Hawk had to remain in the kennel for fifteen days for observation; it was the law. It was the same number of days we had spent so happily on Nantucket. My husband spoke to Fred. You should talk to Fred, he said. When I called, I got Lynn. She spoke to me in a sort of lighthearted way.

She seemed grateful that I had held on to Hawk during the attack. I was too confused by this comment to reply to it. She said, After you left he attacked the noose, but then he calmed down in the cage after we washed the blood off of him. He ate some food. Some dogs get a taste for biting, she said, after they start to bite. Everything she said was wrong.

Finally, she said, He seems to be in conflict. The word seemed to reassure her, it gave her confidence. I couldn't understand a thing she was saying. I wanted someone to tell me why my beloved dog had attacked me so savagely and how I could save both of us. He's just in a lot of conflict now, the girl said. Maybe he had some separation anxiety. He seemed all right for a while after we washed the blood off him. I don't know what to tell you.

Finally, Fred got on the line. He's just not the same dog, Fred said. I know that dog, this isn't him. When I had the noose on him, he was attacking the pole and looking right at me. There was no fear in his eyes, there was nothing in his eyes. I'm no doctor, Fred said, but I think it's a brain tumor. I think something just kicked on or clicked off in him, and you'll never know when it will happen again.

I said, He was a perfectly healthy, happy, loving dog.

This isn't your dog here now, Fred said.

I couldn't bear to call Fred every day. I called him every other day. He has good days and bad days, Fred said. Sometimes you can walk right up to the cage and he just looks at you or he doesn't even bother to look at you. Other times he flings himself at the chain link, attacking it, trying to get at you. Some days he's a monster.

I thought of Hawk's patience, of his happiness, of his dear, grave face. Sometimes, when he slept, he would whimper and his legs would move as though he were walking quickly in a dream. What do you think he's dreaming about, I would ask my husband. Then I would call his name, Hawk, Hawk, it's all right, and he would open one startled eye and look at me and sigh, and then he would be calm again. I couldn't bear the thought of him waiting in the kennel for me to pick him up. I was not going to pick him up. I was going to have him put down, put to sleep, euthanized, destroyed. My love would be murdered. I would murder my love.

The days dragged on. Fred said, He's unreliable. I have no doubt that if you told him to do something he didn't want to do, he would attack you. Anything could set him off, he could turn on anybody. If you slipped and fell, if you were in a helpless position, he could kill you, I have no doubt of it. That's a tough dog. Fred fancied German shepherds and had several of his own whom he exhibited in shows. He's not the same dog anymore, Fred said.

I did not really believe this, that he was not the same dog. I did not think that he had a brain tumor. I thought that something unspeakable and impossible and calamitous had happened to Hawk and me. My husband said, You have to remember him the way he was. If you just dwell on this, if this is all you remember from all the wonderful times you had with him, then shame on you. My husband said, I love him too, I miss him, but I'm not going to mention him every time I think of him. You can talk about him all you want and I'll talk with you, but I'm not going to bring him up again, it makes you too upset.

Upset? I said.

On the fifteenth day, Fred would put a soporific in Hawk's food, and then the vet would arrive and give him a lethal injection. His brain would die and his heart would follow. It would take ten seconds. So often I had sat with Hawk while he ate. He would eat for a while and then pick up a toy and walk around the room with it and then eat some more. Oh, that's so good, I would say to him while he ate. Isn't that good? Oh, it's delicious....

Fred said, I know this is difficult. If he had been run over by a truck, it would be a different matter. You would grieve for him. This is a harder grief.

If I talked about something else at home or if I ate something or if I had a martini again, if I took the time to make a martini rather than just slosh some gin in a glass, my husband said—You seem a little better.

I tried to imagine that Hawk was attempting to reach me telepathically during these days. I went to all his places, for they were my places too, and tried to listen, but nothing was coming through. I didn't expect his apologies of course. For my part I forgave him, but I was going to have him murdered too. We had loved one another and we would never meet again. He never came to me in dreams. I was granted nothing, not the smallest sign.

We had to go to the vet to sign the paper authorizing euthanization. The vet's name was Dr. Turco. There had been Dr. Franks and Dr. Crane and Dr. Yang in my life in the last days, and now there was Dr. Turco. In the parking lot there was a young man with a pit bull in the back of his pickup truck. He was fumbling with the dog's leash somewhat, and it was taking me awhile to get out of our car with my hand in the cast and my aching, crippling malady, my mysterious malady, whatever the hell it was. I passed the dog, sturdy and panting, cute in his ugliness, white and pink with dashes of black about him, a dog with his own charms. Hi there, I said to the dog. The young man seemed unfriendly, he did not seem as nice as his dog. They followed my husband and me into the vet's waiting room, the dog sliding and scrambling across the waxed floor, his nails clicking.

My mind said, *The vet may have an explanation for what happened, an answer. Perhaps some anecdotes at the very least will bring you peace.* Dr. Turco said, Fred tells me that Hawk has become quite dangerous.

I said, It was an aberration, a moment's madness seized him. Could it be a brain tumor?

The vet paused, It's possible ... he said, indicating that it wasn't very likely. He said, So sad. My sympathy and respect for your decision.

It's unusual, isn't it? I asked. For a dog to attack his owner?

It's quite unusual, the vet said. I've never known a dog to attack its owner. Excuse me for just one moment.

He left the room. My God, I said to my husband, did you hear that! He didn't say that, my husband said in anguish. He did! He just did! I said. I'll ask him when he comes back, my husband said.

I've never known personally of it happening, the vet said, in the course of my practice. I'm sure it's probably happened. I'm so sorry.

I signed the paper with my left hand. My signature looked totally unfamiliar to me. Above it, printed by some other hand, was Hawk's name and breed and age and weight. As we returned to the parking lot, the young man we had seen with the pit bull was coming back to his truck from the rear of the vet's office. He was cradling a black garbage bag in his arms, his lips pressed to it. He placed it in the back of the pickup, got into the cab, and sat there for a moment. Then he rubbed his eyes and drove away.

On the sixteenth day, my husband went to the kennel to pay for Hawk's residency there and to pick up his leash. Then he went to the vet and paid for the euthanization and for the cremation that hadn't happened yet. He brought home Hawk's lavender collar from the vet with his tags on it and the St. Francis "Protect Us" medal. I said, That's not Hawk's leash. I wanted to bury Hawk's leash with his ashes and his toys, but I wanted to keep his collar with all the photographs I had of him. That's his leash, my husband said. They bleached it to get the blood out.

Silver Trails is a pet motel, but it also has a crematorium and a cemetery where the pictures of beloved pets, made weatherproof in a silvering process, are mounted on a curved tile wall. The wall was supposed to be capable of withstanding freezing temperatures, but it has not and some of the tiles are cracked. All the dogs shown have been "good" and "faithful." The wall is in a fragrant pine grove, and on the pathway to it there is a plaque, which the owners of Silver Trails are very pleased with. It says, "If Christ Had A Little Dog, It Would Have Followed Him To The Cross." There is no devotion, it is known, like a dog's devotion. Dogs excel in love.

Hawk had been taken from Red Rock to the vet's, but it would be several more days before he was brought to Silver Trails. Actually, only living dogs come to the place so named. Dead dogs come to Trail's End.

I was waiting for someone to call me and say, Your animal will be ready after four, which, when the day arrived, is what they would say. Hawk still did not come to me in dreams. I dreamed instead about worrying that I had not told my mother, who lived only in my dreams. She would feel so badly about Hawk. Surely, I must have told her, I reasoned, but I had forgotten if indeed I had. I wasn't sure. Awake, everywhere I looked, I thought Hawk should be there. He should be here with me. How strange it all is, how wrong, that he is not here. My mind said, *He wants to come back, he wants to come back to his home and be with you, but he can't because you killed him, you had him killed. . . .* My body was my malady, my tedious non-life-threatening banal malady, but my mind was like Job's wife, whose only advice to him was to curse God and die. I felt that I wanted to die.

I was utterly unhappy and when, according to Kierkegaard, one becomes utterly unhappy and realizes the absolute woefulness of life, when one can say and mean it, Life for me has no value, that is when one can make a bid for Christianity, that is when one can begin. One must become crucified to a paradox. One must give up reason.

I listened to Kathleen Ferrier sing from *Orfeo ed Euridice* in her unearthly contralto.

> *What is life to me without thee?*
> *What is left if thou art dead?*
> *What is life, life without thee?*
> *What is life without my Love?*

In the myth of the great musician, Orpheus played music that was so exquisite that not only his fellow mortals but even the wild beasts were soothed and comforted by it. When his Eurydice died, he sang his grief for her to all who breathed the upper air, but he was not able to call her back so he decided to seek her among the dead. It ended badly, of course, though not typically so.

The lovely Kathleen died when she was forty-one years old. Glenn died when he was fifty-one. My mind said, *You haven't done much with your life, think of what those two could have done if they had lived on, you couldn't keep your own pet from tearing you apart, or what do they call them now, not pet, companion animal. . . .*

There was no consolation. Hawk had been my consolation. When the phone rang, a woman's voice said, Your animal will be ready after four. I arrived at Silver Trails, and I was directed to a building with a not unsubtle smokestack. I was told to speak with Michael. But Michael was not there. Michael? I called. I could hear a lawn mower in the distance, and over the sounds of the lawn mower were the sounds of the live dogs barking.

I walked into the building, which had two rooms, then a larger room, open like a garage. There was a stubby, tunnel-like object there, the crematorium oven. There were twenty filled black garbage bags secured with twine on a table and a large sleek golden dog lying free. He was a big dog, lying with his face away from me. He looked fit and not old. One of his ears was folded back on itself in a soft, sad way. I walked outside and just stood there. I didn't know what to do with myself anymore. Eventually, the lawn mower grew closer with the boy named Michael on it. I've come for my dog's remains, I said. His name is Hawk. The boy led me into the building, but he closed the door to the big room. He drew back a curtain that ran along a wall and there were dozens of small black paper shopping bags on the shelves, a bag of a size that might contain something lovely, special, from a boutique. There was a label on each bag that said "TRAIL'S END," and it had the name of a dog and then the owner's name. Inside the bag was a blue-and-white tin with a vaguely Oriental motif of blue swallow-like birds flying. The boy and I searched the shelves for the proper bag. Here he is, I said. The boy pointed to another bag. There's another Hawk, he said. He had a strange, half-smiling grimace. There was grass in his hair and grass stuck to his T-shirt. This is my Hawk, I said. There's my name too. I gestured at the shelves. So many! I said. There's so many!

Oh, sometimes all four shelves are full, the boy said.

At home, I sat on the porch and with great difficulty pried open the lid of the tin with its foolish scene. I used a knife around it. There was cotton on the top and beneath it was a clear bag of ground bones. Hawk's ashes weighed more than those of my mother or my father. We all end up alone, don't we, honey? I said.

And then, in time, my little dream.

Hawk and I are walking among a crowd in near darkness. I am a little concerned for him because I want him to be good. He can hardly move among the people in the crowd, but he pays them no attention. He is close to me, he is calm, utterly familiar, he is my handsome boy, my good boy, my love. Then, of course, I realize that these are the dead and we are both newly among them.

Autumn

THERE IS NO SUCH THING AS TIME GOING STRAIGHT ON TO new things. This is an illusion. Okay? And clinging to this illusion makes it difficult to understand oneself and one's life and what is happening to one. Time is repetition, a circle. This is obvious. Day and night, the seasons, tell us this. Even so, we don't believe it. Time is not a circle, we think. Spring screams the opposite to us, of course, and summer seduces us into believing that we're all going to live forever. Winter couldn't care less what we think about time. But fall cares. Instructive, tactful, subtle, fall is a philosophy all its own. Occult, secretive, taking pleasure in sleep, in rest. Fall's comfortless, honest rot. In the beginning in most places it's showy, the better to mask its melancholy: raging leaves and spanking breezes, edgy with the real cold. And that special, solemn light. For fall is for melancholics and those in love. The torchy sort of love. Forget spring. Spring is nothing but promise, a reproach to melancholics. Spring makes us forget the deal, whereas fall is the deal. The unutterable, unalterable deal.

Fall is. It always comes round, with its lovely patience. If in the beginning it's restless, at the end it's resigned, complete in its waiting, complete in the utter correctness of what it has to tell us. Which is that we're transitory. We're transient, we're temporary, we're all only sometime. We will pass and someone else will take our place. Our pursuit of living founders each time we remember this. Fall is the darkening window, the one Hart Crane had in mind in his poem "Fear," the window on which licks the night.

Why I Write

It's become fashionable these days to say that the writer writes because he is not whole: he has a wound, he writes to heal it. But who cares if the writer is not whole? Of course the writer is not whole, or even particularly well. There's something unwholesome and self-destructive about the entire writing process. Writers are like eremites or anchorites— natural-born eremites or anchorites—who seem puzzled as to why they went up the pole or into the cave in the first place. Why am I so isolate in this strange place? Why is my sweat being sold as elixir? And how have I become so enmeshed with words, mere words, phantoms?

Writers when they're writing live in a spooky, clamorous silence, a state somewhat like the advanced stages of prayer but without prayer's calming benefits. A writer turns his back on the day and the night and tries, like some half-witted demiurge, to fashion other days and nights with words. It's absurd. Oh, it's silly, dangerous work indeed.

A writer starts out, I think, wanting to be a transfiguring agent, and ends up usually just making contact, contact with other human beings. This, unsurprisingly, is not enough. (Making contact with the self—healing the wound—is even less satisfactory.) Writers end up writing stories—or rather, stories' shadows—and they're grateful if they can, but it is not enough. Nothing the writer can do is ever enough.

E. M. Forster once told his friend Laurens van der Post that he could not finish a story that he had begun with great hope, for he felt it held much promise, even brilliance, because he did not like the way it would have to end. Van der Post wrote, "The remark for me proved both how natural stories were to him and how acute was his sense of their significance, but at the same time revealed that his awareness was inadequate for the task the story imposed upon it."

I like van der Post's conception of story—that of a stern taskmaster who demands the ultimate in awareness, that indeed is awareness. The significant story possesses more awareness than the writer writing it. The significant story is always greater than the writer writing it. This is the humble absurdity, the disorienting truth, the exhilarating transmutability, this is the koan of writing.

Malcolm Muggeridge wrote in an essay on Jesus, "When a person loses the isolation, the separateness which awareness of the presence of God alone can give, he becomes irretrievably part of a collectivity with only mass communications to shape its hopes, formulate its values and arrange its thinking."

Without the awareness of separateness, one can never be part of the whole, the nothingness that is God. This is the divine absurdity, the koan of faith.

Jean Rhys said that when she was a child she thought that God was a big book. I don't know what she thought when she was no longer a child. She probably wished that she could think of a big book as being God.

A writer's awareness must never be inadequate. Still, it will never be adequate to the greater awareness of the work itself, the work that the writer is trying to write. The writer must not really know what it is that he is learning to know when he writes, which is more than the knowing of it. A writer loves the dark, loves it, but is always fumbling around in the light. The writer is separate from his work, but that's all the writer is—what he writes. A writer must be smart but not too smart. He must be dumb enough to break himself to harness. He must be reckless and patient and daring and dull—for what is duller than writing, trying to write? And he must never care—caring spoils everything. It compromises the work. It shows the writer's hand. The writer is permitted, even expected, to have compassion for his characters, but what are characters? Nothing but mystic symbols, magical emblems, ghosts of the writer's imagination.

The writer doesn't want to disclose or instruct or advocate, he wants to transmute and disturb. He cherishes the mystery, he protects it like a fugitive in his cabin, his cave. He doesn't want to talk it into giving itself up. He would never turn it in to the authorities, the mass mind. The writer is somewhat of a fugitive himself, actually. He wants to escape his time, the obligations of his time, and, by writing, transcend them. The writer does not like to follow orders, not even the orders of his own organizing intellect. The moment a writer knows how to achieve a certain effect, the method must be abandoned. Effects repeated become false, mannered. The writer's style is his doppelgänger, an apparition that the writer must never trust to do his work for him.

When I began writing essays, I developed a certain style for them that was unlike the style of my stories—it was unelusive and strident and brashly one-sided. They were meant to annoy and trouble and polarize, and they made readers, at least the kind of readers who write letters to the editors of magazines, half nuts with rage and disdain. The letter writers frequently mocked my name. Not only didn't they like my way with words, my reasoning, my philosophy, they didn't believe my name. My morbid attitude, my bitter tongue, my anger, denied me the right to such a name, my given name, my gift, signifier of rejoicing, happiness, and delight.

But a writer isn't supposed to make friends with his writing, I don't think.

The writer doesn't trust his enemies, of course, who are wrong about his writing, but he doesn't trust his friends, either, who he hopes are right. The writer trusts nothing he writes—it should be too reckless and alive for that, it should be beautiful and menacing and slightly out of control. It should want to live itself somehow. The writer dies—he can die before he dies, it happens all the time, he dies as a writer—but the work wants to live.

Language accepts the writer as its host, it feeds off the writer, it makes him a husk. There is something uncanny about good writing—uncanny the singing that comes from certain husks. The writer is never nourished by his own work; it is never satisfying to him. The work is a stranger, it shuns him a little, for the writer is really something of a fool, so engaged in his disengagement, so self-conscious, so eager to serve something greater, which is the writing. Or which could be the writing if only the writer is good enough. The work stands a little apart from the writer, it doesn't want to go down with him when he stumbles or fails or retreats. The writer must do all this alone, in secret, in drudgery, in confusion, awkwardly, one word at a time.

The writer is an exhibitionist, and yet he is private. He wants you to admire his fasting, his art. He wants your attention, he doesn't want you to know he exists. The reality of his life is meaningless, why should you, the reader, care? You don't care. He drinks, he loves unwisely, he's happy, he's sick, it doesn't matter. You just want the work—the Other—this other thing. You don't really care how he does it. Why he does it.

The good piece of writing startles the reader back into Life. The work—this Other, this other thing—this false life that is even less than the seeming of this lived life, is more than the lived life, too. It is so unreal, so precise, so alarming, really. Good writing never soothes or comforts. It is no prescription, neither is it diversionary, although it can and should enchant while it explodes in the reader's face. Whenever the writer writes, it's always three or four or five o'clock in the morning in his head. Those horrid hours are the writer's days and nights when he is writing. The writer doesn't write for the reader. He doesn't write for himself, either. He writes to serve . . . something. Somethingness. The somethingness that is sheltered by the wings of nothingness—those exquisite, enveloping, protecting wings.

There is a little tale about man's fate, and this is the way it is put. A man is being pursued by a raging elephant and takes refuge in a tree at the edge of a fearsome abyss. Two mice, one black and one white, are gnawing at the roots of the tree, and at the bottom of the abyss is a dragon with parted jaws. The man looks above and sees a little honey trickling down the tree, and he begins to lick it up and forgets his perilous situation. But the mice gnaw through the tree and the man falls down and the elephant seizes him and hurls him over to the dragon. Now, that elephant is the image of death, which pursues men, and the tree is this transitory existence, and the mice are the days and the nights, and the honey is the sweetness of the passing world, and the savor of the passing world diverts mankind. So the days and nights are accomplished and death seizes him and the dragon swallows him down into hell and this is the life of man.

This little tale with its broad and beasty strokes seems to approximate man's dilemma quite charmingly, with the caveat that it also applies to the ladies ("she" being "he" throughout here, the writer's woes not limited by gender; like Flannery O'Connor's Misfit, the writer knows there's no enjoyment to be had in this life). This is the story, then, pretty much the story, with considerable latitude to be had in describing those mice, those terrifying mice. But it is not for the writer to have any part in providing the honey—the passing world does that. The writer can't do better than that. What the writer wants to be is the consciousness of the story, he doesn't want to be part of the distraction; to distract is ignoble, to distract is to admit defeat, to serve a lesser god. The story is not a simple one. It is syncretistic and strange and unhappy, and it all must be told beautifully, even the horrible parts, particularly the horrible parts. The telling of the story can never end, not because the writer doesn't like the way it must end but because there is no end to the awareness of the story, which the writer has only the dimmest, most fragmentary knowledge of.

Why do I write? Writing has never given me any pleasure. I am not being disingenuous here. It's not a matter of being on excellent terms with my characters, having a swell time with them, finding their surprising remarks prescient or amusing. I am not on excellent terms with my characters. Rewriting, the attention to detail, the depth of involvement required, the achievement, and acknowledgment of the prowess and stamina and luck involved—all these should give their pleasures, I suppose, but they are sophisticated pleasures that elude me. Writing has never been fun for me. I am too wary about writing to enjoy it. It has never fulfilled me (nor have I fulfilled it). Writing has never done anyone or anything any good at all, as far as I can tell. In the months before my mother died, and she was so sick and at home, a home that meant everything and nothing to her now, she said that she would lie awake through the nights and plan the things she would do during the day when it came—she would walk the dog and buy some more pansies, and she would make herself a nice little breakfast, something that would taste good, a poached egg and some toast—and then the day would come and she could do none of these things, she could not even get out the broom and sweep a little. She was in such depression and such pain and she would cry, If I could do a little sweeping, just that. . . . To sweep with a good broom, a lovely thing, such a simple, satisfying thing, and she yearned to do it and could not. And her daughter, the writer, who would be the good broom quick in her hands if only she were able, could not help her in any way. Nothing the daughter, the writer, had ever written or could ever write could help my mother who had named me.

Why does the writer write? The writer writes to serve—hopelessly he writes in the hope that he might serve—not himself and not others, but that great cold elemental grace that knows us.

Made in the USA
San Bernardino, CA
15 December 2016